Mathematics for Construction Students

C. W. Schofield B.A., B.Sc., F.I.M.A., F.C.S.I.
Principal, Highlands College

Edward Arnold
A division of Hodder & Stoughton
LONDON NEW YORK MELBOURNE AUCKLAND

First published in Great Britain 1961 as *Building Mechanics*

Reprinted 1966
SI edition 1970 Reprinted 1971
Third Edition 1975
Reprinted 1976, 1980, 1983, 1985, 1989, 1990, 1993

ISBN 0 7131 3333 3

Printed and bound in Great Britain for Edward Arnold,
a division of Hodder and Stoughton Limited,
Mill Road, Dunton Green, Sevenoaks, Kent TN13 2YA
by Athenaeum Press Ltd, Newcastle upon Tyne

City College Norwich

Customer name: MR Oliver Hinchliffe
Customer ID: 3932**

Title: Mathematics for Construction Students (3 ed.)
ID: A145893
Due: 25 Oct 2013

Total items: 1
04/10/2013 12:13
Checked out: 1
Overdue: 0
Hold requests: 0
Ready for pickup: 0

Thank you for using the
3M SelfCheck™ System.

Preface

As in previous editions of this text, the content has been divided into short sections complete with graded exercises. Additional revision questions at the end of each chapter provide practice on questions of examination standard. The worked examples have been specially selected both to illustrate the text and to give a clear indication of the kind of presentation which merits full marks in examinations.

C. W. SCHOFIELD

Acknowledgments

The author wishes to express his sincere thanks to the technical examining bodies who have given permission for the reproduction of examination questions. In some cases the units in certain questions have been modified with the full consent of the authorities concerned, but such modifications, and the answers to all the questions, are solely the responsibility of the author.

The abbreviations used represent the following bodies:

D.D.C.T.	Derby and District College of Technology.
E.M.E.U.	East Midland Educational Union.
K.C.E.A.B.	Kent County Examinations and Advisory Board.
N.C.T.E.C.	Northern Counties Technical Examinations Council.
U.E.I.	Union of Educational Institutes.
U.L.C.I.	Union of Lancashire and Cheshire Institutes.
W.J.E.C.	Welsh Joint Education Committee.

Contents

1 Formulae

In many of our calculations we have to use formulae to save the work from becoming unnecessarily laborious, but unfortunately the particular formula we may need to use for a certain problem sometimes has to be changed round first. The rules for this transposition of formulae are basically the same as the rules for solving simple equations, so let us look at these first.

Addition and subtraction

It is quite obvious that if $x + 3 = 5$ then $x = 2$, but how do we arrive at this answer? Logically we should proceed as follows:

write down the original equation	$x + 3 = 5$
subtract three from each side	$x + 3 - 3 = 5 - 3$
which simplifies to	$x = 2$

Now the same result could have been obtained simply by transferring the $+ 3$ to the other side of the equation and changing its sign to $- 3$, but the justification for this lies in the process outlined above.

Let us consider the equation $y - 1 = 7$ comparing the two methods:

$y - 1 = 7$	$y - 1 = 7$
Add 1 to each side,	Take the $- 1$ across and change its sign,
$y - 1 + 1 = 7 + 1$	$y = 7 + 1$
$\therefore \quad y = 8$	$\therefore \quad y = 8$

The working for the second method is shorter, and for this reason is the method usually adopted.

With an equation such as $3z - 4 = 2z + 1$ the idea is to move all the z terms to one side and all the numbers to the other:

$$3z - 4 = 2z + 1$$

$$3z - 2z = 1 + 4$$

$$\therefore \quad z = 5$$

Multiplication and division

Multiplication signs are usually omitted wherever possible. Brackets are used where more than one item is to be multiplied by the same quantity.

Thus with

$$5(2x - 1) - 6 = 9$$

first isolate the term containing x by transferring the 6 to the other side, giving

$$5(2x - 1) = 9 + 6 = 15$$

We now want to remove the bracket, which can be done in either of two ways:

Method I. Multiply out.

$$5(2x - 1) = 15$$
$$10x - 5 = 15$$
$$10x = 15 + 5$$
$$x = \frac{20}{10}$$
$$\mathbf{x = 2}$$

Method II. Isolate the bracket.

$$5(2x - 1) = 15 \quad \text{Divide by 5}$$
$$(2x - 1) = 3$$

The bracket is now unnecessary and may be omitted.

$$2x - 1 = 3$$
$$2x = 4$$
$$x = 2$$

Division is usually indicated by a horizontal straight line, but it must always be remembered that such a line has the same effect as a bracket. It is unnecessary to write $\dfrac{(y + 2)}{3}$ as $\dfrac{y + 2}{3}$ means exactly the same. Thus $\frac{3}{8}x - \frac{5}{8}$ can be put as $\frac{1}{8}(3x - 5)$ or as $\dfrac{3x - 5}{8}$.

This is particularly important where there is a minus sign before a fractional term.

Thus

$$-\frac{x + 3}{4}$$

is the same as $-\frac{1}{4}(x + 3)$ i.e. $-\frac{1}{4}x - \frac{3}{4}$.

With an equation involving fractional terms, it is usually best to clear the fractions first by multiplying through by the lowest common multiple (L.C.M.) of the denominators.

Example:

$$\frac{x + 3}{4} = x - \frac{2x - 1}{3}$$

Multiply through by 12,

$$3(x + 3) = 12x - 4(2x - 1)$$

Remove the brackets,

$$3x + 9 = 12x - 8x + 4$$
$$3x + 9 = 4x + 4$$
$$x = 5$$

Check: L.H.S. $\dfrac{x + 3}{4} = \dfrac{5 + 3}{4} = \dfrac{8}{4} = 2$

R.H.S. $x - \dfrac{2x - 1}{3} = 5 - \dfrac{10 - 1}{3} = 5 - 3 = 2$

Since the value $x = 5$ makes the L.H.S. equal the R.H.S. it is obviously the value which satisfies this equation.

The idea of 'cross-multiplying' is a useful one, but can be used only when each side of the equation consists of a single fractional term. Considering again our example above,

$$\frac{x+3}{4} = x - \frac{2x-1}{3}$$

we must first bring the R.H.S. over a common denominator,

$$\frac{x+3}{4} = \frac{3x-(2x-1)}{3}$$

$$\frac{x+3}{4} = \frac{x+1}{3}$$

Now cross-multiply
$$3(x+3) = 4(x+1)$$
$$3x + 9 = 4x + 4$$
$$x = 5$$

Simultaneous linear equations

Two equations are required for two unknowns, three for three unknowns, etc., these being reduced to an equation in one unknown by elimination or substitution.

Elimination

Consider the two simultaneous equations:

$$3a - 2b = 5 \quad \ldots \quad (1)$$
$$a + 2b = 7 \quad \ldots \quad (2)$$

Adding,
$$4a + 0 = 12$$

b has thus been eliminated, leaving an equation in a alone.

$$\therefore \quad a = \frac{12}{4} = 3$$

Alternatively, multiply equation (2) by 3 to make the coefficient of a the same in each equation:

$$3a + 6b = 21$$

equation (1)
$$3a - 2b = 5$$

Subtracting,
$$0 + 8b = 16$$

an equation involving only b.

$$\therefore \quad b = \frac{16}{8} = 2$$

Substitution

Taking the same equations:

$$3a - 2b = 5 \quad \ldots \quad (1)$$
$$a + 2b = 7 \quad \ldots \quad (2)$$

from equation (2), $a = 7 - 2b$ and this expression for a can be substituted in equation (1) to give:

$$3(7 - 2b) - 2b = 5$$

Multiplying out,

$$21 - 6b - 2b = 5$$
$$8b = 16$$
$$\therefore \quad b = 2$$

Similarly, from equation (1), $2b = 3a - 5$ which can be substituted in equation (2) to give:

$$a + (3a - 5) = 7$$
$$4a - 5 = 7$$
$$a = 3$$

Although it is possible to solve such simultaneous equations entirely by elimination or entirely by substitution, it is usually quicker to use a combination of the two methods.

Example:

Two quantities x and y are connected by the relationship $y = mx + c$, where m and c are constants. Given that $y = 4$ when $x = 3$ and that $y = 2$ when $x = -2$, find the values of m and c.

Putting in the values gives the two equations:

$$4 = 3m + c \quad . \quad . \quad . \quad . \quad (1)$$
$$2 = -2m + c \quad . \quad . \quad . \quad . \quad (2)$$

Subtracting to eliminate c,

$$2 = 5m$$
$$m = 0.4$$

Substitute this value in equation (1),

$$4 = 1.2 + c$$
$$c = 2.8$$

After substituting in one equation it is always a good check to ensure that the values obtained do satisfy the other equation.
e.g. for equation (2)

$$\text{R.H.S. is } c - 2m = 2.8 - 0.8 = 2 = \text{L.H.S.}$$

The methods of elimination and substitution may be applied to any number of simultaneous linear equations.

Powers and roots

We shall sometimes need to take a square or cube root, or get rid of one, but this is quite straightforward; the only point to watch is that a square root can be either positive or negative.

(a) $x^2 = 9$

$x = \pm 3$

i.e. x equals plus 3 or minus 3.

(b) $\sqrt{x - 3} = 2$

$x - 3 = 4$

$x = 7$

Application to formulae

Whereas with simple equations the correct numerical answer must always be the same, the answer to a transposition of formula may be expressed in a variety of equivalent forms. The best is the simplest.

Example 1 To find the breadth of a rectangle given its length and perimeter.

Method I

$$P = 2(l + b)$$

Divide by 2,

$$\frac{P}{2} = l + b$$

$$\therefore \quad b = \frac{P}{2} - l$$

Method II

$$P = 2(l + b)$$

Multiply out

$$P = 2l + 2b$$

$$2b = P - 2l$$

$$\therefore \quad b = \frac{P - 2l}{2}$$

Now it is obvious that the two expressions obtained for b are equivalent, but they give us alternative ways of finding the answer we seek. Method I tells us that we can find the breadth by subtracting the length from half the perimeter, whereas Method II gives the breadth when twice the length is subtracted from the perimeter and the result halved.

Example 2 To find the radius of a sphere of given volume.

$$V = \frac{4}{3}\pi r^3$$

$$4\pi r^3 = 3V$$

$$r^3 = \frac{3V}{4\pi}$$

$$r = \sqrt[3]{\frac{3V}{4\pi}}$$

Example 3 The formula $I = \dfrac{bd^3}{12} + bdh^2$ is used in certain structural calculations. To make b the subject of this formula,

$$I = \frac{bd^3}{12} + bdh^2$$

Multiply by 12,

$$12I = bd^3 + 12bdh^2$$

Take out b as a factor,

$$b(d^3 + 12dh^2) = 12I$$

Divide by the bracket,

$$b = \frac{12I}{(d^3 + 12dh^2)}$$

EXERCISE 1(a)

Solve the following equations:

1. $2x - 3 = x + 2$
2. $3(y + 1) - 2(y + 3) = 1$
3. $\frac{1}{4}(z - 1) = \frac{1}{2}(4 - z)$
4. $4(s + 3) - 2(s - 1) = 4 - 3s$
5. $3(r + 2) - 2r(4 - r) = r(2r - 3)$
6. $2t(t + 2) + t(3 - t) = t^2 - 7$
7. $\dfrac{2p + 3}{5} + p = \dfrac{7p - 1}{3}$
8. $\dfrac{5}{q} - \dfrac{3}{q} = \dfrac{1}{2}$
9. $\dfrac{3}{x + 4} + \dfrac{x}{2} = \dfrac{x - 3}{2}$
10. $\dfrac{x + 1}{x - 2} = \dfrac{x - 3}{x + 4}$
11. $3\sqrt{x + 2} = 6$
12. $2\sqrt{2x + 5} = \sqrt{2 - x}$
13. $\dfrac{3}{\sqrt{7 - 4x}} = \dfrac{1}{\sqrt{3 + 4x}}$
14. $2 + \dfrac{x^3}{4} = \dfrac{x^3}{2}$
15. $x(x - 2) = 2(8 - x)$
16. $(x + 3)(2x - 1) - (x + 1)(x + 4) = 2$
17. $\dfrac{2x + 1}{x - 1} = \dfrac{3x - 4}{x - 4}$
18. $x^2(3 + x^2) = 3(x^2 + 5) + 1$
19. $\dfrac{5}{4 - x} - \dfrac{4 + x}{2} = \dfrac{1}{x - 4}$
20. $\dfrac{7}{x - 1} + x = (x + 1)^2$
21. The volume of a cone is given by $V = \frac{1}{3}\pi r^2 h$. Make r the subject of this formula.
22. The relationship between the principal invested and the interest gained is given by the formula
$$I = \frac{Prn}{100}$$
for simple interest calculations.

    Change the subject to r.
23. The surface area of a cone is given by the formula $S = 2\pi r(r + h)$. Transform for h.
24. From the formula:
$$I = \frac{E}{\sqrt{R^2 + \omega^2 l^2}}$$
obtain an expression for R in terms of the other quantities.
25. The sum of the first n terms of an arithmetic progression is given by the formula:
$$S = \tfrac{1}{2}n[2a + (n - 1)d]$$
Rearrange this formula to give an expression for d.
26. The area of a sector of a circle is given by
$$A = \frac{\pi r^2 \theta}{360}$$
where θ is the angle of the sector measured in degrees. Express this formula with r as the subject.
27. Find an expression for the diameter of a sphere in terms of the volume, given that
$$V = \frac{4}{3}\pi r^3$$

Indices

In the number a^n the letter a is the **base** and n is the **index.** The rules for dealing with indices may be summarized as follows:

(1) Multiplying: When multiplying numbers which are powers of the same base, add the indices.

$$a^n \times a^m = a^{n+m}$$

(2) Dividing: When dividing numbers which are powers of the same base, subtract the index of the denominator from the index of the numerator.

$$\frac{a^n}{a^m} = a^{n-m}$$

(3) Powers of Powers: When a power of any base is raised to a further power, the index numbers are multiplied.

$$(a^n)^m = a^{nm}$$

From these rules arise the following points worth noting:

(i) Any base to the power 0 is equal to 1. $a^0 = 1$

(ii) Fractional indices indicate roots. $\sqrt{a} = a^{\frac{1}{2}}$

(iii) Negative indices indicate reciprocals. $a^{-n} = \dfrac{1}{a^n}$

Using these rules, any number can be expressed as a power of any other number.

Examples:

(1) As powers of 2; $8 = 2^3$ $\frac{1}{4} = 2^{-2}$ $3 = 2^{1.585}$

(2) As powers of 8; $64 = 8^2$ $2 = 8^{\frac{1}{3}}$ $\frac{1}{8} = 8^{-1}$

(3) As powers of 10; $1000 = 10^3$ $0.001 = 10^{-3}$ $20 = 10^{1.301}$

Logarithms

When a number is expressed as a power of a certain base, the index is known as the logarithm of the number to that base. Thus with the previous examples:

(1) To base 2; $\log_2 8 = 3$ $\log_2 \frac{1}{4} = -2$ or $\bar{2}$ $\log_2 3 = 1.585$

(2) To base 8; $\log_8 64 = 2$ $\log_8 2 = \frac{1}{3}$ $\log_8 \frac{1}{8} = -1$ or $\bar{1}$

(3) To base 10; $\log_{10} 1000 = 3$ $\log_{10} 0.001 = -3$ or $\bar{3}$ $\log_{10} 20 = 1.301$

Note that 2 is actually 2^1, and thus $\log_2 2 = 1$. This is true for any base, i.e. $\log_a a = 1$. For most purposes logarithms to the base 10 are used and these are known as *common logarithms.* The tables most frequently used give the values of common logarithms to four decimal places, but it should always be remembered that the fourth decimal place cannot be relied upon, so that answers to calculations should only be given to a number of figures that may reasonably be expected to be accurate.

Equations involving logarithms

If the unknown appears as an index, see if it is possible to take logarithms of each side of the equation.

Examples:

(1) If $2^x = 16$

$2^x = 2^4$

$\therefore x = 4$

taking logarithms, $x \log 2 = \log 16$

$$x = \frac{\log 16}{\log 2}$$

$$x = \frac{1.204\ 12}{0.301\ 03}$$

$$x = 4$$

(2) If $3^x = 5$

taking logarithms, $x \log 3 = \log 5$

$$x = \frac{\log 5}{\log 3} = \frac{0.698\ 97}{0.477\ 12}$$

$$x = 1.465$$

(3) If $100^{2-7x} = 12$

taking logarithms, $(2 - 7x) \log 100 = \log 12$

$$(2 - 7x)2 = 1.0792$$

$$2 - 7x = 0.5391$$

$$7x = 2 - 0.5391 = 1.4609$$

$$x = 0.2087$$

When the equation contains plus or minus signs this method cannot be used since it is impossible to add or subtract by logarithms. This type of equation can usually be factorised. The following example illustrates the factorisation method.

$$4^x + 2^{2x-3} = 9$$

$$2^{2x} + \frac{2^{2x}}{2^3} = 9$$

$$2^{2x}(1 + \tfrac{1}{8}) = 9$$

$$2^{2x} = 8$$

$$2^{2x} = 2^3$$

$$\therefore 2x = 3$$

$$x = 1.5$$

If the unknown appears as a logarithm, put back into index form.

Examples:

(1) $\log_{10} x = 3$

$x = 10^3$

$x = 1000$

(2) $2 \log_{10} x = 3$

$\log_{10} x^2 = 3$

$x^2 = 10^3$

$x = \sqrt{1000} = 10\sqrt{10}$

$x = 31.62$

(3) $\log_{10} \frac{3}{4}x = 5$

$\log_{10} \frac{3}{4} + \log_{10} x = 5$

$\log_{10} x = 5 - \log_{10} 0.75$

$\log_{10} x = 5 - \bar{1}.875\ 06$

$\log_{10} x = 5 - (-1 + 0.8750)$

$\log_{10} x = 5.124\ 94$

$x = 133\ 333$

alternatively,

$\frac{3}{4}x = 10^5$

$3x = 4 \times 10^5$

$x = 133\ 333$

(4) $3 \log_4 x = 5$

$$\log_4 x = \frac{5}{3}$$

$$x = 4^{\frac{5}{3}}$$

Taking logarithms to base 10

$$\log_{10} x = \frac{5}{3} \log_{10} 4$$

$$\log_{10} x = \frac{5}{3} \times 0.6021$$

$$\log_{10} x = 1.0035$$

$$x = 10.08$$

The last example illustrates a change of base, and this is sometimes necessary. Let us examine a straightforward change from one base to another.

To change $\log_a x$ to base b.

Let $y = \log_a x$

then $a^y = x$

taking logs to base b

$$y . \log_b a = \log_b x$$

$$\therefore\ y = \frac{\log_b x}{\log_b a}$$

Now let $\dfrac{1}{\log_b a} = z$

$$\log_b a = \frac{1}{z}$$

$$a = b^{1/z}$$

$$a^z = b$$

$$z . \log_a a = \log_a b$$

$$z = \log_a b$$

hence $\dfrac{1}{\log_b a} = \log_a b$

Therefore a change of base is given by

$$\log_a x = \frac{\log_b x}{\log_b a} \text{ or } \log_b x . \log_a b$$

As stated previously, logarithms are indices, and they must therefore obey the laws of indices for multiplication, division and powers.

(1) *Multiplication:* When multiplying, add the logs.

$$\log xy = \log x + \log y$$

(2) *Division:* When dividing, subtract the logs.

$$\log \frac{x}{y} = \log x - \log y$$

(3) *Powers:* When a number is raised to a power, the log of the number must be multiplied by the power.

$$\log x^n = n . \log x$$

From these rules and the formula for changing the base of a logarithm the following important points arise:

(i) The log of any number to itself as base is 1.

$$\log_a a = 1$$

(ii) The log of one to any base is zero.

$$\log_a 1 = 0$$

(iii) The log of any number is equal to minus the logarithm of its reciprocal.

$$\log_a y = -\log_a\left(\frac{1}{y}\right)$$

and conversely,

$$\log\left(\frac{1}{y}\right) = -\log y$$

The slide rule

The slide rule works on the same laws that have been outlined for indices and logarithms. The numbers on the scales are marked out at distances corresponding to the logarithms of the numbers. The A and B scales are identical. To multiply two numbers together, find the first on the A scale and place against it the mark 1 (one) on the movable B scale. Now look for the second number along the B scale and read off the value against it on the A scale. This is the required product, which has actually been obtained by adding together the appropriate logarithms and reading off the answer as a number.

The C and D scales are another identical pair, but with the numbers spaced twice as far apart as on the A and B scales. Squares and square roots are thus found by referring straight across the rule on A and D scales. Slide rules may be obtained with a variety of other useful scales upon them, but the A, B, C and D scales are basic and appear on almost all slide rules.

EXERCISE 1(b)

1. Simplify:
   (a) $\dfrac{6^3 \times 4^2}{3^2 \times 8^3}$ (b) $\dfrac{y^3 \times (2y)^4}{8y^5}$ (c) $\dfrac{6^x \times 2^{x+1}}{4^x \times 3^{x-2}}$

2. Simplify:
   (a) $[\frac{81}{16}]^{\frac{1}{4}}$ (b) $[1\frac{9}{16}]^{-\frac{1}{2}}$ (c) $[3\frac{3}{8}]^{-\frac{2}{3}}$

3. Express the following numbers as powers of 2:
   (a) $\frac{1}{8}$ (b) $\sqrt[3]{32}$ (c) 20 (d) $\frac{5}{8}$

4. Express the following numbers as powers of 10:
   (a) 0.001 (b) 0.003 (c) 0.01 (d) $\frac{6}{7}$

5. Evaluate:
   (a) $\log_2 16$ (b) $\log_9 27$ (c) $\log_4 \frac{1}{8}$

6. Find the values of:
   (a) $\log_e 6$ (b) $\log_e 4.35$ (c) $\log_e 0.36$

7. Solve the equations:
   (a) $4^x = 512$ (b) $(\frac{1}{4})^x = 8$ (c) $9^{x+1} + 3^{2x-1} = 28$

8. Solve the equations:
   (a) $3^{2x+1} = 243$ (b) $128^{1/x} = 4$ (c) $2^{3x+1} - 8^{x-1} = 15$

9. Find the value of x to 3 significant figures:
   (a) when $3^{2+x} = 5^{x-1}$ (b) when $6^{x+\frac{1}{2}} = 5^{x+\frac{3}{4}}$

10. Solve the equations:
    (a) $2 \log_{10} x = 3$ (b) $\log_{10} \frac{1}{3}x = 0.45$

11. Find the value of x to 3 significant figures, when
    (a) $\log_2 3x = 1.4$ (b) $\log_5 \frac{7}{8}x = 2$ (c) $\log_3 x^2 = \frac{3}{8}$

12. Simplify:
    (a) $\dfrac{\log 8 - \log \frac{1}{2}}{2 \log 2}$ (b) $\dfrac{\log 6 + \log \frac{1}{4}}{\log 3 - \log 2}$

13. Evaluate:
    (a) $\dfrac{1 - \log_{10} 80}{1 - \log_{10} 40}$ (b) $\dfrac{\log_2 80 + 2 \log_2 \frac{1}{4}}{\log_2 100 - 2}$

14. Evaluate:
    (a) $\log_6 8.5$ (b) $\log_7 1.394$ (c) $\log_{\frac{1}{2}} 0.28$

Evaluations from formulae

It is usually much better to work with the letters of the formula as long as possible before substituting the given values.

Example: If a lead sphere of radius 30 mm could be beaten into the shape of a uniform disc of radius 60 mm, what would be the thickness of the disc?

Let t be the thickness of the disc.

Volume of disc $= \pi R^2 t$

$$\text{Volume of sphere} = \frac{4}{3}\pi r^3$$

$$\therefore \quad \pi R^2 t = \frac{4}{3}\pi r^3$$

$$R^2 t = \frac{4}{3} r^3$$

$$t = \frac{4r^3}{3R^2}$$

$$t = \frac{4 \times 30 \times 30 \times 30}{3 \times 60 \times 60}$$

$$t = 10 \text{ mm}$$

Expressions involving square roots can sometimes be simplified.

Method I. By factors. To resolve $\sqrt{1323}$.

$$1323 = 3.3.3.7.7 = 3^3.7^2$$

There is an even power of 7, so the square root of this part can be taken straightaway. The power of 3 is odd, therefore upon dividing the index by 2 for the square root there will be a remainder of a single 3 left under the square root.

$$\sqrt{1323} = \sqrt{3^3.7^2}$$

$$\sqrt{1323} = 3.7.\sqrt{3}$$

$$\therefore \quad \sqrt{1323} = 21\sqrt{3}$$

Method II. By rationalisation.

(1) $\cos 45^\circ = \dfrac{1}{\sqrt{2}}$ Multiply top and bottom of this fraction by $\sqrt{2}$,

$$= \frac{1}{\sqrt{2}} \times \frac{\sqrt{2}}{\sqrt{2}}$$

$= \dfrac{\sqrt{2}}{2}$ which is easier to evaluate.

(2) Given that $\sin 75^\circ = \dfrac{\sqrt{3}+1}{2\sqrt{2}}$

and $\cos 75^\circ = \dfrac{\sqrt{3}-1}{2\sqrt{2}}$

find tan 75°.

Now $\tan 75^\circ = \dfrac{\sin 75^\circ}{\cos 75^\circ}$

$$\therefore \quad \tan 75^\circ = \frac{\sqrt{3}+1}{\sqrt{3}-1}$$

Now the difference of two squares $a^2 - b^2 = (a + b)(a - b)$ so to eliminate the square root from the denominator, we multiply top and bottom by the conjugate factor, in this case $\sqrt{3} + 1$.

$$\tan 75^\circ = \frac{\sqrt{3}+1}{\sqrt{3}-1} \times \frac{\sqrt{3}+1}{\sqrt{3}+1}$$

$$\tan 75^\circ = \frac{3 + 2\sqrt{3} + 1}{3 - 1}$$

$$\therefore \quad \tan 75^\circ = 2 + \sqrt{3}$$

Note: $(\sqrt{x} + \sqrt{y})^2$ is *not* equal to $x + y$, but just as

$$(a + b)^2 = a^2 + 2ab + b^2$$

so

$$(\sqrt{x} + \sqrt{y})^2 = x + 2\sqrt{xy} + y$$

Approximations

When evaluating a formula it is always advisable to work out an approximate answer first as a rough check. When the calculations are done on a slide rule it is absolutely essential that an approximation to the result should be worked out in order to fix the position of the decimal point.

The following approximations are useful for this purpose:

$$\pi \simeq \frac{22}{7} \qquad \pi^2 \simeq 10 \qquad \sqrt{2} \simeq 1.4 \qquad \sqrt{3} \simeq 1.7 \qquad \sqrt{5} \simeq 2\tfrac{1}{4}$$

Examples:

(1) $$\frac{9.8 \times 3\sqrt{3}}{4.08 \times \sqrt{5}} \simeq \frac{10 \times 3 \times 1.7}{4 \times 2\frac{1}{4}} \simeq \frac{50}{9} \simeq 5\tfrac{1}{2}$$

Approximate answer $5\frac{1}{2}$. Answer by logarithms 5.582.

(2) A semicircular groove is cut round a pillar, the volume of material removed being given by:

$$V = \pi^2 r^2 \left[R - \frac{4r}{3\pi}\right]$$

Calculate V when $R = 12$ and $r = 1\frac{1}{2}$.

Approximately,

$$V = 10(1\tfrac{1}{2})^2 \left[12 - \frac{6}{3\pi}\right]$$

$$V \simeq 10 \times 2\tfrac{1}{4} \left[12 - \frac{7}{11}\right]$$

$$V \simeq 25 \times 12$$

$$V \simeq 300$$

Accurately,
$$V = \pi^2 \times 2\tfrac{1}{4}\left[12 - \frac{6}{3\pi}\right]$$
$$= \pi^2 \times 2\tfrac{1}{4}(12 - 0.6366)$$
$$= 252.4$$

Now let us look into this a little closer: it is obvious that the figure 300 is sufficiently accurate to enable the decimal point to be inserted after a slide rule calculation, but not near enough to provide a guide to the true answer. Let us go back to the line

$$V \simeq 10 \times 2\tfrac{1}{4}\left[12 - \frac{7}{11}\right]$$

For convenience and speed we put 25 for $10 \times 2\tfrac{1}{4}$, and ignored the fraction $\tfrac{7}{11}$. Both of these approximations had the effect of increasing the value of V, and it would have been better to have made them compensate for one another, one to increase, the other to decrease, the value of V.

Also, 22 or 23 would have been a closer approximation to $10 \times 2\tfrac{1}{4}$ than 25. Therefore, combining these, it would have been nearer to have put

$$V \simeq 22 \times 12 \text{ or } 23 \times 11$$

giving
$$V \simeq 264 \text{ or } 243$$

Both of which are much closer to the value 252.4.

Accuracy

Answers are frequently given containing more figures than are justified. If, for example, the dimensions of a rectangular building plot were measured as 21.4 m × 68.7 m, long multiplication would give the area as 1470.18 m^2. Since the original measurements were only accurate to three figures, the derived area cannot be more accurate than this and therefore the area should be given as 1470 m^2.

If measurements were taken of a rectangular block of stone and the dimensions found to be 428 mm × 172 mm × 86 mm, then the product of these three figures would be 6 330 976 and it might appear reasonable to give the volume of the block as 6331 cm^3. In practice, such a block could not be measured with absolute accuracy and it would be reasonable to expect that each measurement could vary by a millimetre either way. The actual accuracy of measurement is dependent upon the uniformity of the block, the accuracy of the scales on the rule and the skill of the person using it. Suppose the errors are all in the same direction, i.e. all three measurements are either too large or too small. Possible dimensions could vary from 427 mm × 171 mm × 85 mm to 429 mm × 173 mm × 87 mm. Hence the volume should lie between 6206 cm^3 and 6457 cm^3. To give the volume as 6331 cm^3 is misleading since it indicates an accuracy greater than can be justified. (As in the previous example, a calculated answer cannot be accurate to more significant figures than the original dimensions.) It would be more reasonable to correct the 6331 to only two significant figures and give the volume as 6300 cm³.

Whereas for most practical purposes it is sufficient to reduce the final answer to the least number of significant figures given in the prime data, some scientific investigations benefit from an indication of the probable accuracy. An alternative way of giving a reasonable figure for the volume for this purpose would be 6330 cm³ (± 130 m³). The probable margin of error thus revealed can also be put as a percentage and this would give 6330 cm³ (± 2%) as a reasonable estimate.

To ensure that measurements are taken with the smallest possible error:

(i) Use a good rule marked out accurately.

(ii) Use a rule of material unaffected by change of temperature.

(iii) Use a thin rule, or a transparent one with the markings on the underside, so that the divisions are as close as possible to the surface being measured. Do not view from an angle.

(iv) Measure at several points along the length and take the average.

EXERCISE 1(c)

1. The volume of a frustum of a symmetrical pyramid is given by

$$V = \tfrac{1}{3}h(A + B + \sqrt{AB})$$

Given that $V = 218$ when $A = 25$ and $B = 49$, find h.

2. A cylinder of solid metal has a diameter D and length l. A hollow cylinder of the same volume has a diameter D externally, and d internally, its length being L. Find an expression for the ratio $\dfrac{L}{l}$ and evaluate when $d = \tfrac{1}{2}D$.

3. Without using tables or slide rule, find the values of:

(a) $\sqrt{512}$ (b) $\sqrt{4563}$ (c) $\sqrt{2178}$ (d) $\sqrt{6\tfrac{1}{8}}$

given that $\sqrt{2} = 1.414$ and $\sqrt{3} = 1.732$.

4. Simplify:

(a) $\dfrac{1 - \sqrt{2}}{1 + \sqrt{2}}$ (b) $\dfrac{\sqrt{3} + \sqrt{5}}{\sqrt{3} - \sqrt{5}}$ (c) $\dfrac{\sqrt{20} + 2}{1 - \sqrt{5}}$

5. $\sin 15^\circ = \dfrac{\sqrt{3} - 1}{2\sqrt{2}}$, $\cos 15^\circ = \dfrac{\sqrt{3} + 1}{2\sqrt{2}}$, show that $\tan 15^\circ = 2 - \sqrt{3}$.

6. If the radius of a circle were measured incorrectly as 10.2 m instead of 10 m, show that the error in the calculated area would be 4.04%.

7. A student's measurement of the radius of a cylinder as 63 mm was 4% too small, and his measurement of the height as 204 mm was 2% too large. Calculate the percentage error these figures would give on the volume.

REVISION EXERCISE 1

1. Solve $\dfrac{x - 1}{2} - \dfrac{x - 2}{3} + \dfrac{x - 3}{4} = \dfrac{2}{3}$ (U.L.C.I.)

2. Solve $\dfrac{x}{4} - \dfrac{5x + 8}{6} = \dfrac{2x - 9}{3}$ (U.L.C.I.)

3. Solve $\dfrac{3x + 2}{5(7x - 1)} = \dfrac{1}{6}$ (U.L.C.I.)

4. Solve $\frac{1}{2}(x + 2) - \frac{1}{3}(x + 1) = \frac{1}{4}(x + 6) - 1$. (U.L.C.I.)

5. Solve the equation
$$5(x - 1) - 4(x - 2) = 2(4x - 9)$$ (U.L.C.I.)

6. Solve the equation:
$$5\left(\frac{x}{2} + \frac{3}{4}\right) - 2\left(\frac{x}{4} - \frac{1}{8}\right) = \frac{1}{2}(7x - 16)$$ (U.L.C.I.)

7. (a) Solve the equation $\dfrac{a+2}{4} + \dfrac{2a-3}{6} = \dfrac{a+3}{3}$

   (b) Simplify $7a^2 - a(b + 2a - c) + 5ac$ (U.E.I.)

8. Solve the simultaneous equations:
$$3x + 4y = 13$$
$$2x - 3y = 20$$ (K.C.E.A.B.)

9. Solve for x and y:
$$3y = 9x - 33$$
$$3x = 1 - 4y$$ (N.C.T.E.C.)

10. (a) Given $C = \dfrac{E}{R + r}$, find r in terms of the other quantities.

    (b) The equation for the deflection of a cantilever loaded at the free end with a weight W is
$$y = \frac{W}{EI}\left(\tfrac{1}{2}lx^2 - \tfrac{1}{6}x^3\right)$$ (U.L.C.I.)

    Express l in terms of the other quantities.

11. The following formula applies to thin tubes under external pressure

$$P = \frac{m^2}{m^2 - 1} 2E \left(\frac{t}{D}\right)^3$$

Express t in terms of the other quantities. (U.L.C.I.,

12. Rearrange the following formula for Young's modulus so that L is the subject of the equation instead of Y:

$$Y = \frac{3Wl^2(2D + L)}{2ab^3S}$$ (U.L.C.I.)

13. In this equation express K in terms of the other symbols:

$$\frac{A}{a} = \frac{Tu}{v}\sqrt{\frac{K}{k}}$$ (U.L.C.I.)

14. Evaluate, using logarithms,

$$\frac{\sqrt[4]{(0.0235)^3} \times (11.13)^{\frac{1}{2}}}{(14.76)^{2/5}}$$ (U.E.I.)

15. (a) Using logarithms, evaluate:

$$\frac{2.741}{0.327} + \sqrt{\frac{174.2}{4.658}}$$

(b) Calculate the percentage error in taking 10.0 as an approximation for $(3.142)^2$. (W.J.E.C.)

16. Evaluate, without the use of logarithms or a slide rule:

(a) $\frac{6.1275 \times 0.032}{0.000\,24}$ (b) $\frac{0.008\,64 \times 2.1875}{0.003 \times 0.09}$

(c) $(203.125 \div 0.0007)0.003\,36$. (U.L.C.I.)

17. Evaluate by logarithms:

(a) $\frac{7.460 \times 82.90 \times \sqrt{23.45}}{0.0724 \times \sqrt{0.8675}}$ (b) $\frac{15.89 \times \sqrt{101\,400} \times 0.2002}{3.142 \times \sqrt{0.6246}}$

(U.L.C.I.)

18. A formula used in reinforced concrete work is:

$$\frac{bn}{2A} = \frac{d - n}{n} \cdot \frac{E}{E_c}$$

(a) Rearrange this formula to obtain an expression for d.

(b) Evaluate n when $b = 10$, $d = 20$, $A = 1.41$, and $\frac{E}{E_c} = 15$. (N.C.T.E.C.)

19. If f = crushing strength of the material, l = length of the strut, r = radius of gyration of the cross-sectional area of the strut, P = buckling stress, c = a constant depending upon the material and the nature of the ends of the strut, then the Rankine–Gordon formula for the buckling load is

$$P = \frac{f}{1 + c\left(\frac{l^2}{r^2}\right)}$$

Find r when $f = 28$, $c = \frac{1}{1500}$, $l = 15$, and $P = 15.7$. (U.L.C.I.)

20. (a) In stress calculations for concentrated loads in girders the following formula is used

$$W_2 = \frac{\frac{W}{n}}{1 + \frac{md}{z}}$$

Find m in terms of the other quantities.

(b) In the formula:

$$N = \sqrt{\frac{(15D)^5 H}{L}}$$

without using logarithms or slide rule, calculate N if $D = 1.6$, $H = 50$, and $L = 3$. (U.E.I.)

21. The formula

$$af = \frac{h + (n + 1)P}{2} - \sqrt{\left[\frac{h + (n + 1)P}{2}\right]^2 - hP}$$

is used to calculate the safe working stress f tons per in^2 for a strut. Using logarithms, calculate f if $a = 2.36$, $h = 18$, $n = 0.2152$ and $P = 24.93$.

(U.L.C.I.)

22. (a) Given the formula: $P = Q\sqrt{(A - x^2)}$.
Change the subject of the formula to x.

(b) Evaluate by logarithms:

(i) $\dfrac{1}{\sqrt[3]{0.3276}}$ (ii) $(5.864 \times 0.0728)^{\frac{3}{5}} - (0.6285)^{\frac{1}{3}}$. (U.E.I.)

23. (a) Use logarithms to evaluate:

$$\frac{(0.247)^2 \times 24.76}{\sqrt{0.4562}}$$

(b) Determine the value of T, given that :

$$\frac{T}{t} = \left(\frac{1}{r}\right)^{\alpha - 1}$$

where $\alpha = 1.404$, $t = 521$, and $r = \frac{1}{4}$. (U.L.C.I.)

24. (a) Use logarithms to evaluate:

$$\sqrt[3]{\frac{(4.72)^2 \times 4.56}{0.009\,13 \times (237)^2}}$$

(b) Determine the value of: $(3.412)^{0.62} + 3.713$. (U.E.I.)

25. The Brinell hardness number (n) for a piece of metal is given by the formula:

$$n = \frac{F}{\frac{D}{2}\left(D - \sqrt{D^2 - d^2}\right)}$$

For a piece of steel the Brinell number is 228, when $F = 3000$ and $D = 10$. Calculate d. (U.L.C.I.)

26. From the formula:

$$\left(\frac{P_1}{P_2}\right)^{1/n} = \frac{V_2}{V_1}$$

express P_2 in terms of P_1, V_1, V_2 and n. Evaluate P_2 when $P_1 = 6.2$, $V_2 = 8.64$, $V_1 = 2.16$, and $n = 1.4$. (N.C.T.E.C.)

27. (a) Evaluate by logarithms: $(0.378)^{\frac{2}{3}}$.

(b) The formula:

$$d = \sqrt{\left[\frac{2M}{Kx(1 - \frac{1}{3}x)}\right]}$$

occurs in concrete design.

(i) Evaluate d by logarithms when $M = 200\,000$, $K = 5400$, $x = 0.36$.

(ii) Change the subject of the formula to K. (U.E.I.)

28. Express in simpler form with positive indices only:

$$\left(\frac{2a^{-3}b}{3a^2b^2}\right)^{-2}$$

(N.C.T.E.C.)

29. A formula used in relation to the bending of a beam supported at each end is

$$2t = \sqrt{\frac{8L + 3Mg}{2bdY}}$$

(a) Evaluate t if $d = 0.02$, $b = 0.78$, $Y = 1.36 \times 10^7$
$L = 3$, $M = 0.23$, $g = 32$.

(b) Change the subject of the formula to L. (U.E.I.)

30. Rearrange the formula

$$\frac{D}{d} = \sqrt{\frac{f+p}{f-p}}$$

to obtain a new formula for p.
Evaluate p when $D = 9.54$, $d = 6.23$, and $f = 5.97$. (N.C.T.E.C.)

31. Solve the equations:

(a) $3^x = 436$.

(b) $x = 0.0714^{-1.56}$.

(c) $6^{x+1} = 7^x$. (U.E.I.)

32. (a) Transpose the formula given, to make r the subject:

$$S = 2\pi R[R - \sqrt{(R^2 - r^2)}].$$

(b) Find the value of x which satisfies the equation $5^{x+1} = 8^x$.

(c) Solve the equation $\log_x 0.027 = 3$.

(d) Simplify: $\left(\frac{9x^2}{16y^2}\right)^{-\frac{1}{2}}$ (U.L.C.I.)

2 Graphs

IT IS USUAL to draw two axes at right angles intersecting at the origin O, the horizontal axis OX is often referred to as the x-axis, and the vertical axis OY as the y-axis. The position of any point on the graph is then fixed if its distance from each axis is given, such distances being known as **co-ordinates.** The distance of the point from the x-axis is called its **ordinate.** and its distance from the y-axis is called its **abscissa.**

Plotting a graph

It is advisable to observe the following details:

(1) Give the graph a title to explain what it is all about.

(2) Make full use of the graph paper; the larger the graph the more accurate it should be.

(3) Choose scales which are easy to work from: it is far better to let one unit on the graph paper represent 1, 2, 5 or 10 say, than 3, 7 or 9.

(4) It is often possible to start one or both scales from zero, but never do so if it will result in crowding the points into one small corner of the graph paper.

(5) Write sufficient figures along each axis to help in reading the graph, but avoid overcrowding.

(6) Label each axis to explain what the numbers represent, not forgetting to put in the units where appropriate.

(7) Mark the points as clearly as possible: one method of doing this is to place a very small ink dot in the correct position with a small fine pencil circle round it to make it easy to find. Never use thick ink crosses.

(8) If the graph is a straight line, use a ruler or other straight edge. If a curve has to be drawn, French curves or other devices may be used, but often the curve has to be sketched in freehand. In this case always sketch the curve from the inside with as long a stroke as possible. The ideal is a single fine pencil line through the points, but practice is needed to achieve this, and at first it may be necessary to erase an attempt with gentle strokes of a soft rubber. This is why a small ink dot is preferable when marking the points.

The straight line graph

Any equation which can be expressed in the form $y = mx + c$ will give a straight line graph. The quantity in the position of y in the equation is plotted on the vertical axis, and the other variable along the horizontal axis.

The quantity m represents the gradient or slope of the line, and thus in fig. 1

$$m = \frac{RS}{QS} = \tan \theta$$

whilst c represents the length OP, i.e. the intercept on the y-axis. It follows from this that if $c = O$ the graph passes through the origin, and equations with the same value of m have graphs which are parallel straight lines.

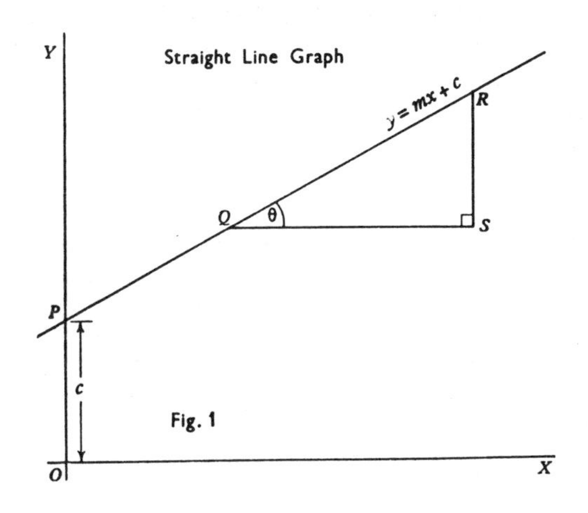

Fig. 1

A typical case of a useful straight line graph is one to illustrate the connection between temperatures on the Fahrenheit and Centigrade scales. The relationship is

$$F = \frac{9}{5}C + 32$$

since 0°C = 32°F and 100°C = 212°F. Temperatures below 32°F will be negative on the Centigrade scale, and in fact, temperatures as low as − 273°C have been obtained. In fig. 2 therefore the axes have been extended to show negative temperatures on both scales. A third point was obtained by finding the Fahrenheit temperature corresponding to − 100°C from the given formula

$$F = \frac{9}{5}C + 32$$

This gave − 100°C = − 148°F. Although only two points are really essential to fix the position of the straight line, it is advisable to plot three or possibly four as a check that all the points are correctly placed.

Interpolation and extrapolation

Intermediate values may now be determined from the graph without repeated evaluations from the same formula. Thus, to convert 75°C to

Fahrenheit, we simply trace the 75°C line up until it crosses the graph, then read across horizontally on to the Fahrenheit axis to get the value 167°F. Similarly, to change − 121°F to Centigrade, trace the level − 121°F across until the graph is reached, then follow up to the value − 85°C. The determination of such intermediate values from the graph is known as **interpolation.** If a value outside the given range is required, this is obtained by **extrapolation,** in which the graph is assumed to continue in the same manner and the value estimated either by continuing the line of the graph as far as necessary, or by calculation, which, in the case of a straight-line graph, could be done by similar triangles.

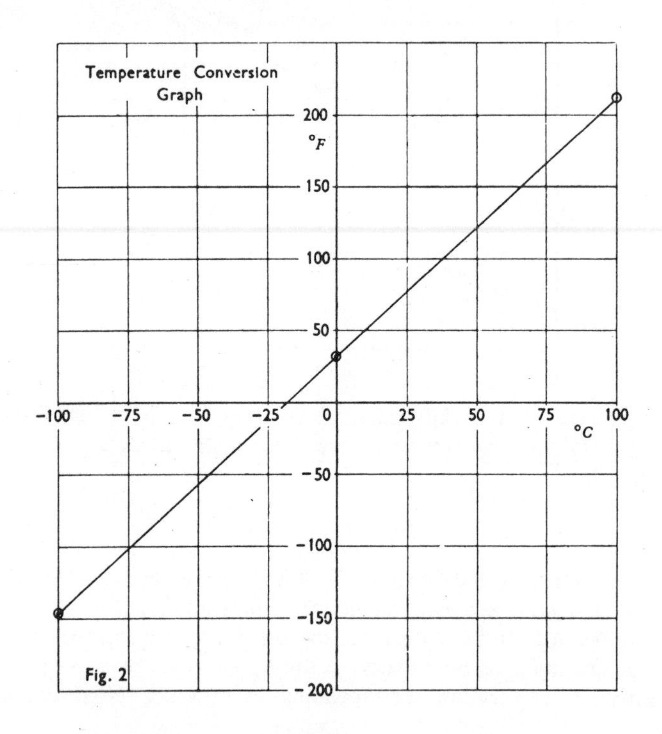

Fig. 2

If the co-ordinates of any two points are given, the equation of the straight line joining those points may be obtained either by plotting the points, joining them with a straight line, and finding its slope and intercept (i.e. m and c), or, alternatively, by putting the two sets of x and y values into the equation $y = mx + c$ and solving the two simultaneous equations so obtained algebraically as in chapter 1. A straight line graph which slopes downwards from left to right will have a negative slope, whilst one which crosses the vertical axis below the origin will have a negative value for c.

Example: Plot the graph of $3x + 2y + 1 = 0$ for values of x from -3 to $+2$.

The given equation is not in the form $y = mx + c$, so it must first be transposed. Thus

$$y = -\tfrac{3}{2}x - \tfrac{1}{2}$$

from which it can be seen that this straight line graph will have a slope of $-\frac{3}{2}$ and an intercept on the y-axis of $-\frac{1}{2}$.

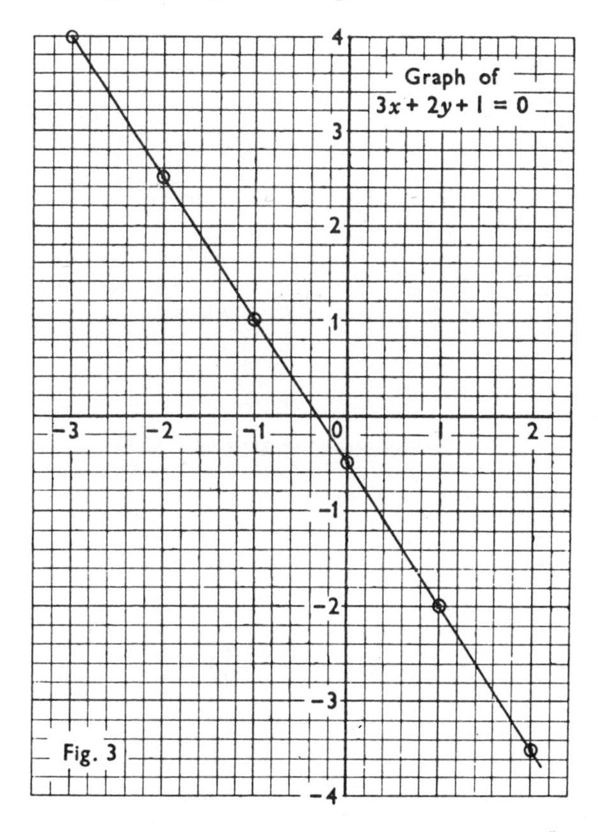

Fig. 3

From the equation, values of y corresponding to the values of x from -3 to $+2$ are determined and set out in a table.

x	-3	-2	-1	0	1	2
y	4	$2\frac{1}{2}$	1	$-\frac{1}{2}$	-2	$-3\frac{1}{2}$

For a graph known to be a straight line three or four of these values would suffice although the positions of them all are shown on fig. 3.

EXERCISE 2(a)

1. Given that 1 metre per second is equivalent to 3.6 kilometres per hour, plot a graph for converting speeds of up to 200 km/h into m/s. From your graph find:
   (a) the speed in m/s equivalent to 150 km/h;
   (b) the speed in km/h equivalent to 39 m/s.

2. Plot a graph of weight against cost for meat at 80 p per kilogramme for weights up to 2 kg. From your graph find:
   (a) the cost of a piece of meat weighing 456 g;
   (b) the weight for £1.05.

3. Find the gradients and intercepts on the y-axis for each of the following without plotting the graphs:
   (a) $2y = 4x + 3$ (b) $y + 5x = 7$ (c) $4y - 6x = 9$
   (d) $2x - 8y = 5$ (e) $9x = 3y - 2$ (f) $7 - x - y = 0$
   (g) $\frac{x+5}{2} = \frac{y-3}{4}$ (h) $\frac{y}{4} - \frac{x}{3} = 1$

4. Plot the graphs of the following equations using only one pair of axes, and thus show that they form a square, and find its area.
   (a) $y = \frac{3(2-x)}{2}$ (b) $\frac{y}{2} = \frac{x+4}{3}$
   (c) $3x + 2y + 3 = 0$ (d) $2x = 3y + 1$

5. Draw the straight line graph through the points (4, 2) and (1, −3). Measure the intercept and gradient and so find the equation of the straight line. Check by substituting the values in the equation $y = mx + c$ and solving the simultaneous equations for m and c.

6. Plot the graphs of:
$$y = 3x - 2$$
$$y = \tfrac{1}{2}x + 2$$
   and find the values of x and y at the point of intersection. Check the result obtained by solving these two simultaneous equations algebraically.

The determination of laws

We have seen that any equation of the form $y = mx + c$ will give a straight line graph, and the converse of this is also true, i.e. that if a given set of values when plotted give points which within the limits of experimental accuracy lie on a straight line, then a relationship of the form $y = mx + c$ must exist.

Example: The following results are from an experiment with a set of pulley blocks. The forces shown are measured in newtons.

Effort E	8	11.9	16	20.2	24	27.9
Load W	20	40	60	80	100	120

It is known that these should be related by the law $E = aW + b$. If this is so, determine the values of the constants a and b.

compare $$E = aW + b$$
with $$y = mx + c$$

The variable E is in the position of y and should therefore be plotted along the vertical axis, whilst W, being in the place of x, will go horizontally. If the resulting graph is a straight line, then the gradient of it will be a and the intercept on the vertical axis will be b.

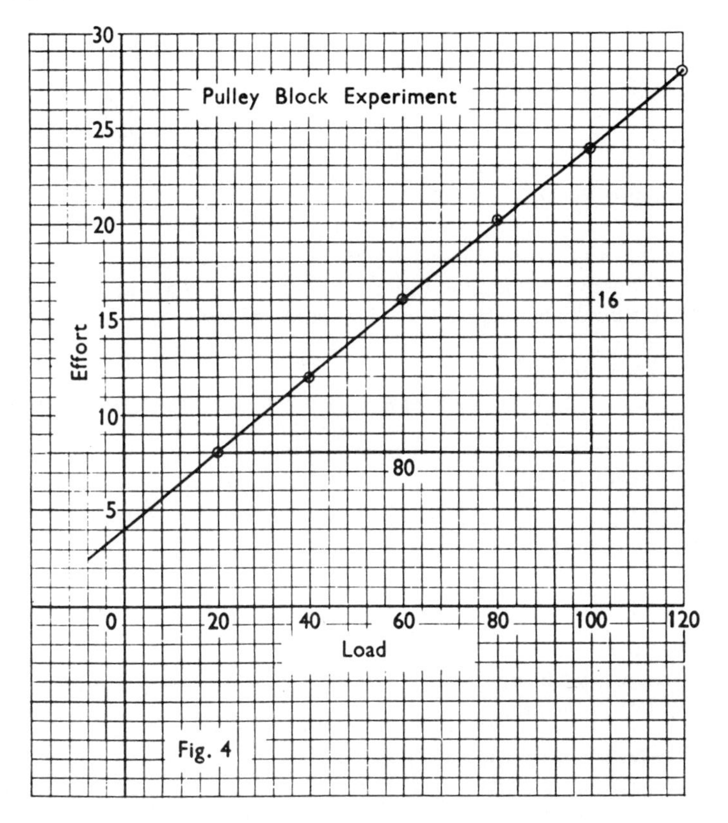

Fig. 4

The graph is shown in fig. 4, and is clearly a straight line. This proves the validity of the relationship $E = aW + b$. The intercept is seen to be 4, therefore $b = 4$. The gradient is $\frac{16}{80} = \frac{1}{5}$, therefore $a = \frac{1}{5}$. Hence the complete relationship is $E = \frac{1}{5}W + 4$.

It is not essential to use the gradient and intercept to determine the constants. The alternative method is to select two convenient points on the line and read off their co-ordinates, then substitute these figures in the equation of the line so as to form two simultaneous equations which may be solved algebraically. Values from the given table should never be used for this purpose unless it has first been determined that the corresponding points do lie exactly on the straight line. Whichever method is adopted, it is always advisable not to select points too close together.

In our example we have shown that the first and fifth points lie exactly on the line. Using the values for these points:

$E = 8$ when $W = 20$ and $E = 24$ when $W = 100$ we obtain the two equations:

$$8 = 20a + b \qquad . \quad . \quad . \quad . \quad (1)$$

$$24 = 100a + b \qquad . \quad . \quad . \quad . \quad (2)$$

Subtracting (1) from (2),

$$16 = 80a \qquad \therefore \quad a = \frac{16}{80} = \frac{1}{5}$$

Substituting in (1),

$$8 = 4 + b \qquad \therefore \quad b = 4$$

These values for a and b being exactly the same as before.

Now, however, let us investigate the position if, say, the second and fourth sets of values are used: $E = 11.9$ when $W = 40$, and $E = 20.2$ when $W = 80$.

The equations are:

$$11.9 = 40a + b \qquad . \quad . \quad . \quad . \quad (1)$$

$$20.2 = 80a + b \qquad . \quad . \quad . \quad . \quad (2)$$

Subtracting (1) from (2),

$$8.3 = 40a \qquad \therefore \quad a = 0.2075$$

Substituting in (1),

$$11.9 = 8.3 + b$$

$$b = 3.6$$

These values for a and b show a variation of as much as 10% from the true values, which is the result of selecting points which are both slightly off the line of the graph and too close together.

Laws reducible to straight line form

The equation $y = ax^2 + c$ would give a curve if we plotted y against x, and although the intercept in this case would enable us to find the value of the constant c, the gradient is not constant and would not give us a directly. We can, however, represent any quantities we choose along the axes, and so the method in this case is to plot y against x^2, taking x^2 as our new variable in place of x. Comparison of

$$y = mx + c$$

and

$$y = ax^2 + c$$

shows that plotting y against x^2 will give a straight line graph, gradient a and intercept c.

Many other laws can be reduced to straight line forms by similar methods and thus the values of the constants can be determined provided that sets of values of the variables are given. Methods of dealing with the most common types of equations are as follows:

(1) $y = ax^3 + b$, etc.

With equations of the forms $y = ax^2 + b$, $y = ax^3 + b$, $y = a\sqrt{x} + b$, $y = \frac{a}{x} + b$ and $y = \frac{a}{x^2} + b$, plot y against x^2, x^3, $\sqrt{x}$, $\frac{1}{x}$ or $\frac{1}{x^2}$ as appropriate. Comparison with $y = mx + c$ shows that the gradient m will give a, and the intercept c will give the value of b.

(2) $y = ax^n$

Take logs of each side:

$$\log y = \log a + n.\log x$$

Compare $\qquad y = \qquad c + mx$

Plot $\log y$ against $\log x$ to obtain a straight line graph with gradient n and intercept $\log a$; a can then be found by using a table of antilogarithms.

(3) $y = ab^x$

Take logs of each side:

$$\log y = \log a + x.\log b$$

Compare $\qquad y = \qquad c + xm$

Plot $\log y$ against x to obtain a straight line graph with gradient $\log b$ and intercept $\log a$; use antilogs to find a and b.

(4) $y = ae^{bx}$

Take logs of each side:

$$\log y = \log a + bx.\log e \quad (\log_{10}e = 0.4343)$$

Compare $\qquad y = \qquad c + mx$

Plot $\log y$ against $\log x$ to obtain a straight line graph with gradient $0.4343b$ and intercept $\log a$.

(5) $y = ax^2 + bx + c$

This contains three constants and it is necessary to find the value of one of them before attempting to obtain a straight line form. The method is to plot a graph of y against x or, at least, as much of it as is necessary to find the intercept on the y axis. This gives the value of c.

Rearrange the equation:

$$y - c = ax^2 + bx$$

Divide by x:

$$\frac{y-c}{x} = ax + b$$

Draw up a set of values for $\frac{y-c}{x}$ and plot against x. The gradient will be a and the intercept b.

Other equations containing three constants e.g. $y = ax^n + b$ can be dealt with similarly.

If the given values are known to be accurate, they can be substituted in the given law and the resulting equations solved simultaneously. When the given values are liable to experimental error, it is better to plot the graph, select two appropriate points on the curve and use their co-ordinates to substitute to give the simultaneous equations. For a law with three constants three points would be required and three simultaneous equations.

Example: Two variables, x and y, are connected by an equation of the form $y = ax^n$. Given $y = 67.5$ when $x = 3$ and $y = 20$ when $x = 2$, find the value of the constants a and n. (U.L.C.I.)

Taking logs: $\log y = \log a + n \log x$

Substituting values: $\log 67.5 = \log a + n \log 3 \quad . \quad . \quad . \quad (1)$

$\log 20 = \log a + n \log 2 \quad . \quad . \quad . \quad (2)$

The values of the logs can be put in at this stage and the equations worked out with the resulting numbers, or otherwise:

Subtracting: $\log 67.5 - \log 20 = n(\log 3 - \log 2)$

$$n = \frac{\log 3.375}{\log 1.5} = \frac{0.5283}{0.1761}$$

$$\therefore \quad n = 3$$

Substituting in (2) $\log 20 = \log a + 3 \log 2$

$$20 = 8a$$

$$\therefore \quad a = 2.5$$

EXERCISE 2(b)

1. In an experiment, readings are taken of temperature T°C after time S seconds. It is known that T and S are connected by a relationship $T = a + bS$. Given that the temperature is 10°C after 5 seconds, and 20°C after 15 seconds, find the values of the constants a and b.

2. A straight line graph is given by the equation $y = mx + c$. Given that $y = 4$ when $x = 2$, and that $y = 7$ when $x = 8$, find the values of the constants m and c. Hence find the value of x which makes $y = 5$.

3. Corresponding values of two observed quantities are given in the following table:

x	1	2	3	4	5
y	5.1	10.8	21.2	35.1	52.9

It is thought that the quantities follow a law of the form $y = a + bx^2$. Represent these values as a straight line graph and hence find the most probable values of a and b. (U.E.I.)

4. The following table shows the values of the pitch P and the corresponding depth d of a screwthread:

P	1.0	0.8	0.5	0.3	0.2	0.1
d	0.67	0.54	0.34	0.21	0.14	0.07

Find the law connecting d and P by plotting a graph. (U.L.C.I.)

5. The table below shows the cost of ladders of various lengths. Plot a graph of cost against length and use it to find the cost of a ladder 8 m in length.

Cost in £	7	8	10	13	17	22	27
Length in m	2	$2\frac{1}{2}$	$3\frac{1}{2}$	5	7	$9\frac{1}{2}$	12

6. A group of construction engineers have a certain installation initially valued at £12 500. The table below shows the depreciation in the value of this installation. Plot a graph of value against time in years, and from the graph find how many years it will take for the value to drop to $\frac{1}{10}$ of its initial value.

Time in years	0	2	6	10	16
Value in £	12 500	11 250	8750	6250	2500

7. The following results are from an experiment to relate the quantity of a concrete hardener used to the time required for the concrete to set.

Hardener in l/m³	(H)	38	28	22	14	9
Time in hours	(T)	1	$1\frac{1}{2}$	2	$3\frac{1}{2}$	6

It is believed that the formula connecting H and T is of the form $T = kH^{-5/4}$

Verify this by plotting the graph of T against $H^{-5/4}$ and hence determine the value of the constant k.

8. The results of an experiment on the deflection of a beam are given in the table below. Verify that the deflection (y) and the length of the beam (x) are connected by the relationship $y = ax^n$, and use graphical methods to estimate the values of the constants a and n.

x in m	1.6	1.8	2.1	2.5	3.2
y in mm	6.6	10.5	19.5	39.0	104.8

9. The mass M which can be safely distributed uniformly on a girder of span L varies as follows:

M kilograms	184	140	103	68	40
L metres	1.6	2.0	2.5	3.2	4.0

Plot a graph of ML against L^2 and show that this is a straight line. Explain why this verifies the relationship $M = aL + b/L$ and find the values of the constants a and b. (U.E.I.)

REVISION EXERCISE 2

1. A relationship of the form $y = ax^3 + b$ is satisfied by the values $y = 1$ when $x = 2$, $y = 29$ when $x = 4$. Find the values of the constants a and b.

2. A parabola passes through the points ($x = 2$, $y = 13$); ($x = 5$, $y = 100$); and ($x = -1$, $y = -20$). If the equation is of the form $y = ax^2 + bx + c$, determine the value of a, b and c. (N.C.T.E.C.)

3. Find the values of constants A, B and C, so that the curve whose equation is $y = Ax^2 + Bx + C$ shall satisfy the condition that when $x = 0$, 1 and 2, $y = 3$, 0 and 2 respectively. (N.C.T.E.C.)

4. In a test on a piece of wire the following values of load L, and extensions y, were found. (Some of the values are missing.)

L (kg)	3	—	10	13	15	18
y (mm)	0.95	1.25	2.35	2.95	—	3.95

The equation connecting y and L is of the type $y = aL + b$. Find graphically the values of a and b and the missing values in the table. (W.J.E.C.)

5. The effort E required to raise a load W with a builders' winch is given by

$$E = aW + b$$

The following values, for E and W are found experimentally:

E	7.1	9.2	10.0	13.1	14.4	17.5	21.2
W	30	45	50	70	80	100	125

By drawing a suitable graph estimate the most probable values of the constants a and b. (U.E.I.)

6. The effort (E) to raise a load (W) for a lifting tackle is shown in the following table. If E and W are connected by an equation of the form $E = aW + b$, plot a suitable graph and from it find the values of a and b.

Effort	5	8.5	11.5	14.5	18	20.7	24.0	27.0
Load	7	14	21	28	35	42	49	56

(N.C.T.E.C.)

7. The quantity of common salt (w grams) which would dissolve in a fixed volume of water at temperature t°C is given in the table below.

w (g)	27	29	30	31	33	35	37
t (°C)	10	20	25	30	40	50	60

By plotting a graph of w against t show that they are connected by a law of the form $w = mt + c$. From the graph find the values of m and c and the value of w when $t = 35°C$. (U.L.C.I.)

8. The following table shows the relative values of two quantities S and Q. Express the results graphically.

Q	1	1.8	2.6	4	5.2	6.3	8
S	31	28	25	19	14	10	3

From the graph obtain an equation giving S in terms of Q. (U.L.C.I.)

9. When testing a floor for stiffness the following results were obtained giving the deflection (d) at the centre of a panel of the floor with the load (w) used.

w	10	15	30	45	65	80	100
d	0.17	0.21	0.31	0.42	0.56	0.66	0.80

Plot a graph of d against w and from the graph obtain an equation connecting d and w. (U.L.C.I.)

10. The following table shows how the safe, uniformly distributed load on a particular steel girder decreases as the span increases.

Load (L)	45.7	36.3	26.7	20.7	18.7	17
Span (S)	8	10	14	18	20	22

Assuming that the quantities are related by a law of the form $L = \frac{k}{S}$, plot a straight line graph and from the graph find the value of k. (U.L.C.I.)

11. In a heat transmission experiment the following results were obtained for the temperature difference T and the time t.

T	12.6	15.9	22.0	31.6	39.8	56.1	63.1
t	1.0	1.76	4.26	10.0	17.8	40.0	56.2

It is known that the relation between T and t is of the form $T = At^b$. Show that this is so and find the constants A and b. (K.C.E.A.B.)

12. Verify that the following values of x and y satisfy approximately a law of the form $y = ax^n$ and determine values for the constants a and n.

x	10	16	26	39	64	100
y	20	25	32	40	50	63

(W.J.E.C.)

13. The following table gives the corresponding values of V, the volume, and P, the pressure of a mixture in a gas-engine cylinder.

V	0.8	1.5	3.0	6	9
P	200	89	35	12.6	7.2

It is known that a law of the form $PV^n = C$ connects P and V, C being a constant. By drawing a suitable graph show that this is so and find the actual law. Use the graph to find the value of V when $P = 20$. (N.C.T.E.C.)

14. In an experiment, the head of water pressure h and the velocity of flow v were recorded as follows:

h	5	7	10	12	15
v	3.43	4.0	4.8	5.2	5.8

Verify by means of a graph that the relationship between h and v is of the form $v = kh^n$ and find probable values of the constants k and n. (U.L.C.I.)

15. In an experiment to determine the law connecting compressive strength of a cement paste (S) and the water–cement ratio (x) the following results were obtained:

Water–cement ratio x	0.4	0.6	0.8	1.0	1.2	1.4
Compressive strength S	5850	3750	2450	1550	1000	650

Determine an equation of the type $S = A/B^x$ which satisfied these results, where S is the compressive strength at 28 days, A and B are constants, and x is the water–cement ratio. (N.C.T.E.C.)

16. In an investigation into the transmission of sound through panels of various thicknesses, the sound absorption A and the corresponding thickness t were measured and the following results obtained:

A	0.83	1.30	1.70	1.93	2.44	2.90
t	5	8	12	15	25	40

Plot a graph of A against $\log_{10}t$.
Do the results confirm that A and t are related by a formula of the type $A = m \log_{10}t + c$ over the given range?
If so, what are the values of the constants m and c. (U.E.I.)

3 Solution of Equations

Factors

Expressions containing obvious common terms are easy to factorise.

e.g. $$2x^3 + 6x^2 = 2x^2(x + 3)$$

and $$\begin{aligned} ax - ay + bx - by &= a(x - y) + b(x - y) \\ &= (a + b)(x - y) \end{aligned}$$

The method is simply to take the common terms outside a bracket.

Certain other expressions may be factorised by using one of the standard forms:

$$\begin{aligned} a^2 - b^2 &= (a - b)(a + b) \\ a^3 - b^3 &= (a - b)(a^2 + ab + b^2) \\ a^3 + b^3 &= (a + b)(a^2 - ab + b^2) \end{aligned}$$

More difficulty is presented by second degree expressions such as $5x^2 + 7x - 6$ since no common terms are apparent. Once the centre term, $7x$, has been split into $10x - 3x$, the rest is easy, thus:

$$\begin{aligned} 5x^2 + 7x - 6 &= 5x^2 + 10x - 3x - 6 \\ &= 5x(x + 2) - 3(x + 2) \\ &= (5x - 3)(x + 2) \end{aligned}$$

Unfortunately it is not always easy to see how to split the centre term so as to give the required form.

Consider the general form $ax^2 + bx + c$; to factorise this we need to split the centre term bx into two parts, say $px + qx$, such that the product pq is equal to the product ac, and the sum $p + q$ is equal to b.

Applying this rule to the expression $5x^2 + 7x - 6$, we have $pq = -30$ and $p + q = 7$, and from these simple equations we get the values $p = 10$ and $q = -3$ by inspection. Thus the centre term $7x$ must be put as $10x - 3x$ to factorise.

To take another example, let us find the factors of $6x^2 + 11x + 4$. We need to split the centre term $11x$ into two parts px and qx such that $p + q = 11$ and $pq = 24$. By inspection p and q are 8 and 3.

Thus: $$\begin{aligned} 6x^2 + 11x + 4 &= 6x^2 + 8x + 3x + 4 \\ &= 2x(3x + 4) + 1(3x + 4) \\ &= (2x + 1)(3x + 4) \end{aligned}$$

Note that it would not make any difference if p and q were reversed:

$$\begin{aligned} 6x^2 + 11x + 4 &= 6x^2 + 3x + 8x + 4 \\ &= 3x(2x + 1) + 4(2x + 1) \\ &= (3x + 4)(2x + 1) \end{aligned}$$

Methods of solving quadratic equations

Whereas a linear equation contains no power of x higher than the first, a second degree equation contains no power higher than the second, although it may contain first degree terms also. All quadratic equations can be put in the form $ax^2 + bx + c = 0$, but it should be noted that it is possible for b or c to be zero.

Method I—Solution by factors

If a second degree equation can be expressed as the product of two factors then the solutions are obtained by putting each of the factors equal to zero in turn.

Example 1

$$x^2 + 3x = 0$$
$$x(x + 3) = 0$$

therefore either $x = 0$ or $x + 3 = 0$

$$x = 0 \text{ or } -3$$

Example 2

$$x^2 - 16 = 0$$
$$(x - 4)(x + 4) = 0$$
$$\therefore \quad x - 4 = 0 \text{ or } x + 4 = 0$$
$$\therefore \quad x = +4 \text{ or } -4$$

Example 3 $x^2 + x - 12 = 0$

To split the centre term, $p + q = 1$ and $pq = -12$,

which are satisfied by $p = +4$ and $p = -3$.

Hence

$$x^2 + 4x - 3x - 12 = 0$$
$$x(x + 4) - 3(x + 4) = 0$$
$$(x - 3)(x + 4) = 0$$
$$\therefore \quad x - 3 = 0 \text{ or } x + 4 = 0$$
$$\therefore \quad x = 3 \text{ or } -4$$

Example 4 $10x^2 + x - 2 = 0$

For this $p + q = 1$ and $pq = -20$, hence $p = 5$ and $q = -4$.

$$10x^2 + 5x - 4x - 2 = 0$$
$$5x(2x + 1) - 2(2x + 1) = 0$$
$$(5x - 2)(2x + 1) = 0$$
$$\therefore \quad 5x - 2 = 0 \text{ or } 2x + 1 = 0$$
$$\therefore x = 0.4 \text{ or } -0.5$$

Example 5 $x^4 - 13x^2 + 36 = 0$

This is obviously not a second degree equation in x since it contains x^4 but it can be transformed into one by substituting A for x^2.

$$A^2 - 13A + 36 = 0$$

factorising,
$$(A - 4)(A - 9) = 0$$
$$A = 4 \text{ or } A = 9$$
$$x^2 = 4 \text{ or } x^2 = 9$$
$$\therefore \quad x = \pm 2 \text{ or } x = \pm 3$$

Method II—Solution by completing the square

For example 2 in the previous method we had $x^2 - 16 = 0$, and this could also have been solved as follows:

$$x^2 = 16$$

Take the square root of each side,

$$x = \pm 4$$

Now it is possible to do this because the L.H.S. is a perfect square, x^2 and contains no term in x. Suppose we take the equation

$$x^2 + 2x = 15$$

Add 1 to each side to make the left-hand side a perfect square,

$$x^2 + 2x + 1 = 16$$

i.e.
$$(x + 1)^2 = 16$$

Take the square root of each side,

$$x + 1 = \pm 4$$
$$\therefore \quad x = +3 \text{ or } -5$$

In general, to obtain a perfect square on the L.H.S., the rule is to take half the coefficient of x, square it and add to each side.

Example 6 $x^2 - 6x = 16$

The coefficient of the term in x is -6, half of this is -3, squared gives 9. Add 9 to each side:

$$x^2 - 6x + 9 = 25$$
$$(x - 3)^2 = 25$$
$$x - 3 = \pm 5$$
$$\therefore \quad x = 8 \text{ or } -2$$

Example 7 $5x^2 + 4x - 12 = 0$

Rearranging,
$$5x^2 + 4x = 12$$
$$x^2 + \frac{4}{5}x = \frac{12}{5}$$

Half the x coefficient is $\frac{2}{5}$, which squared gives $\frac{4}{25}$,

$$x^2 + \frac{4}{5}x + \frac{4}{25} = \frac{12}{5} + \frac{4}{25}$$

$$\left(x + \frac{2}{5}\right)^2 = \frac{64}{25}$$

$$x + \frac{2}{5} = \pm\frac{8}{5}$$

$$\therefore \quad x = 1.2 \text{ or } -2$$

Example 8 Consider the general quadratic equation

$$ax^2 + bx + c = 0$$

$$ax^2 + bx = -c$$

$$x^2 + \frac{b}{a}x = -\frac{c}{a}$$

$$x^2 + \frac{b}{a}x + \left(\frac{b}{2a}\right)^2 = \frac{b^2}{4a^2} - \frac{c}{a}$$

$$\left(x + \frac{b}{2a}\right)^2 = \frac{b^2 - 4ac}{4a^2}$$

$$x + \frac{b}{2a} = \pm\frac{\sqrt{b^2 - 4ac}}{2a}$$

$$\therefore \quad x = \frac{-b \pm \sqrt{b^2 - 4ac}}{2a}$$

Method III—Solution by formula

Using the formula for the solution of the equation $ax^2 + bx + c = 0$ which has been derived above by the method of completing the square, all valid solutions of all quadratic equations may be found.

Example 9 $\quad 5x^2 + 11x - 12 = 0$

compare $\quad ax^2 + bx + c = 0$

here $a = 5$, $b = 11$, $c = -12$.

Substitute in

$$x = \frac{-b \pm \sqrt{b^2 - 4ac}}{2a}$$

thus

$$x = \frac{-11 \pm \sqrt{121 - 4 \times 5 \times (-12)}}{2 \times 5}$$

$$x = \frac{-11 \pm \sqrt{361}}{10}$$

$$x = \frac{-11 \pm 19}{10}$$

$$x = 0.8 \text{ or } -3$$

Example 10 $\qquad 5x^2 - 20x - 11 = 0$

Compare $\qquad ax^2 + bx + c = 0$

here $a = 5,\ b = -20,\ c = -11$,

$$x = \frac{20 \pm \sqrt{400 + 220}}{10}$$

$$x = \frac{20 \pm 24.90}{10}$$

$$x = 4.49 \text{ or } -0.49$$

Method IV—Solution by graph

The graph of a quadratic equation is a parabola. For the general equation $y = ax^2 + bx + c$ the curve will have a minimum if a is positive, or a maximum if a is negative. The branches of the curve either side of this turning point are always symmetrical. The curve will cross the x-axis in two places if b^2 is greater than $4ac$ and these will give the roots of the equation, i.e. the required solutions. If b^2 and $4ac$ are equal the equation is a perfect square and the curve will then just touch the x-axis at its minimum or maximum value. If b^2 is less than $4ac$ then the curve will never cross the x-axis and the roots are said to be imaginary. If c is zero then the curve will pass through the origin.

Example 11

Plot the curve $y = 2x^2 - 3x - 6$ and hence solve the equations

$$2x^2 - 3x - 6 = 0$$

$$2x^2 - 3x - 5 = 0$$

and $\qquad 2x^2 - 3x - 9 = 0$

The graph of $y = 2x^2 - 3x - 6$ is shown in fig. 5 and is a parabola with a minimum at $x = \frac{3}{4}$, $y = -7\frac{1}{8}$. It crosses the x-axis where $x = -1.14$ and $x = 2.64$, and thus the solutions of the equation $2x^2 - 3x - 6 = 0$ are $x = 2.64$ or -1.14.

Now the equation $2x^2 - 3x - 5 = 0$ can be put as $2x^2 - 3x - 6 = -1$, and so we can obtain the solutions to this equation simply by reading off on the graph of $y = 2x^2 - 3x - 6$ the values of x for which $y = -1$, i.e. $x = -1$ or $2\frac{1}{2}$.

Similarly, $2x^2 - 3x - 9 = 0$ can be put as $2x^2 - 3x - 6 = 3$, so we read off on the graph the values of x when $y = 3$, i.e. $x = -1.5$ or 3.

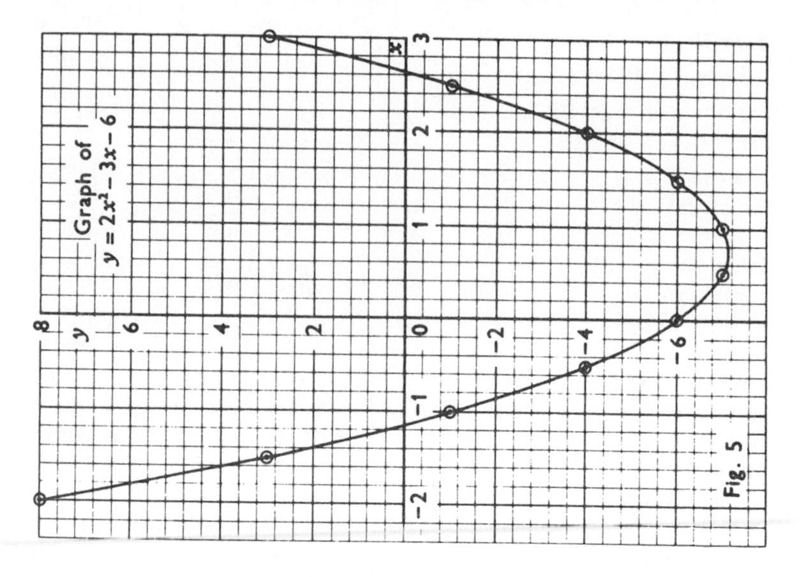
Graph of
$y = 2x^2 - 3x - 6$
y
x
8
6
4
2
0
-2
-4
-6
-2
-1
1
2
3

Fig. 5

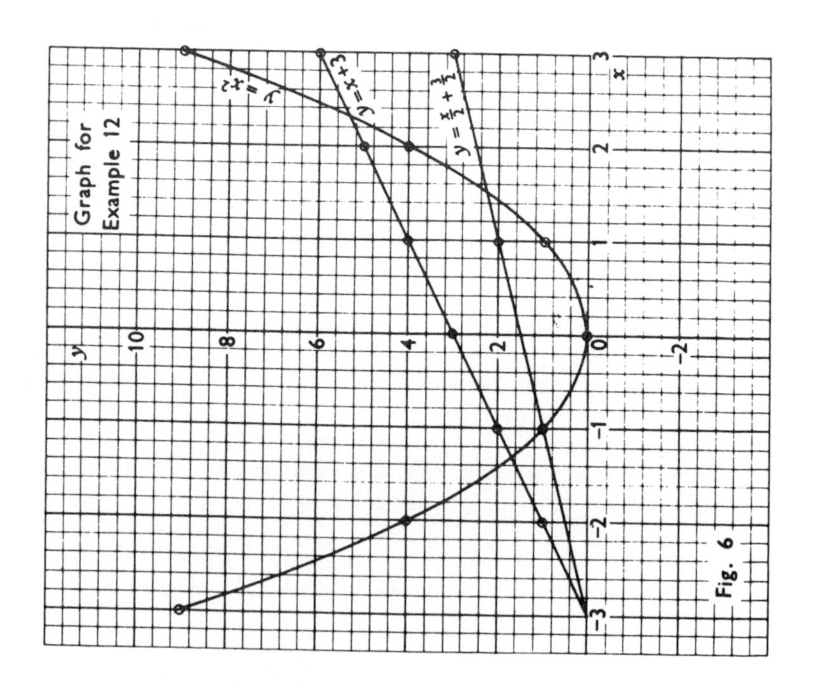
Graph for
Example 12
$y = x^2$
$y = x + 3$
$y = \frac{x}{2} + \frac{3}{2}$
y
x
10
8
6
4
2
0
-2
-3
-2
-1
1
2
3

Fig. 6

Example 12

Draw the graph of $y = x^2$ and use it to solve the equations

$$x^2 = x + 3$$

and $$2x^2 - x - 3 = 0$$

The graph of $y = x^2$ is the parabola shown in fig. 6. It has a minimum at the origin and is symmetrical about the y-axis.

To use this curve to solve the equation $x^2 = x + 3$, we need to find the values for which x^2 and $x + 3$ are the same. We therefore plot the straight line graph $y = x + 3$ and find where it crosses the graph of $y = x^2$. At these points y for each graph is the same, therefore $x^2 = y = x + 3$ at these points, and the x co-ordinates of the points of intersection are the roots of the quadratic equation $x^2 = x + 3$. The line $y = x + 3$ has been drawn on fig. 6, and it crosses the parabola where $x = 2.3$ or -1.3.

For the equation $2x^2 - x - 3 = 0$ the same method may be adopted if the equation is first rearranged into the form $x^2 = \frac{1}{2}x + 1\frac{1}{2}$. The required

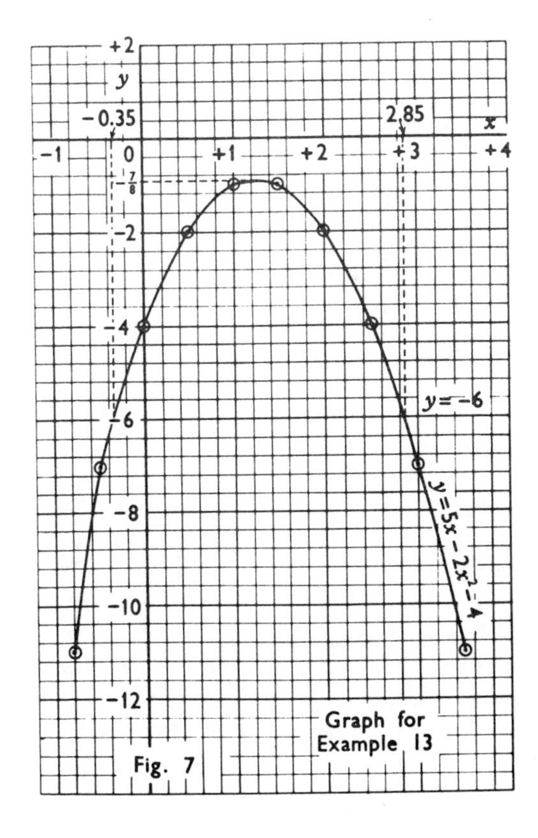

Fig. 7

solutions are then given by the intersections of the straight line $y = \frac{1}{2}x + 1\frac{1}{2}$ with the parabola $y = x^2$, which, from fig. 6 are $x = 1.5$ and -1.0.

Example 13

Draw the graph of $y = 5x - 2x^2 - 4$ and hence show that the equation $5x - 2x^2 - 4 = 0$ has no real roots. Use the graph to solve the equation $5x - 2x^2 + 2 = 0$. Find also from the graph the value of c in the equation $5x - 2x^2 + c = 0$ for this equation to have equal roots.

The graph of $y = 5x - 2x^2 - 4$ is shown in fig. 7. It is a parabola with a maximum since the coefficient of x^2 is negative. It does not cross the x-axis, so it cannot have any real roots.

The equation $5x - 2x^2 + 2 = 0$ may be put as $5x - 2x^2 - 4 = -6$ and its roots are therefore given by the values of x where $y = -6$ on the graph $5x - 2x^2 - 4 = 0$. From fig. 7 the values are 2.85 and -0.35.

The equation $5x - 2x^2 + c = 0$ will have equal roots if it just touches the x-axis at its maximum. Now the graph of $y = 5x - 2x^2 - 4$ in fig. 7 passes below the x-axis at a distance of $\frac{7}{8}$ units, so that the equation required is $5x - 2x^2 - 4 = -\frac{7}{8}$, i.e. $5x - 2x^2 - 3\frac{1}{8} = 0$. The value of the constant c for equal roots is thus $3\frac{1}{8}$.

EXERCISE 3(a)

1. Factorise:
   (a) $2x + 4y - x^2 - 2xy$ (b) $6ab - 2ac + 4bc - 3a^2$
2. Simplify:
   (a) $\dfrac{2x^2 + 2xy}{5xy - 5y^2} \times \dfrac{10x^2 - 10xy}{4x + 4y}$ (b) $\dfrac{1}{x - 2} - \dfrac{1}{x + 2}$
3. Factorise:
   (a) $9 - 4x^2$ (b) $\frac{1}{4}a^2 - b^4$
4. Factorise:
   (a) $x^3 - 8$ (b) $27 + x^3y^6$
5. Factorise and solve the following equations:
   (a) $x^2 - 8x + 15 = 0$ (b) $x^2 = 3x + 28$
   (c) $2x^2 - 3x + 1 = 0$ (d) $2x^2 + 5x = 12$
   (e) $3x^2 + 5 = 8x$ (f) $4x^2 - x = 0$
   (g) $9x = 16x^2$ (h) $3x^2 + 2x = 5$
   (i) $4x^2 + 4x + 1 = 0$ (j) $9x^2 + 4 = 12x$
6. Solve by completing the square:
   (a) $4x^2 + 5x = 6$ (b) $3x^2 + 4x = 6$
7. Use the formula to solve:
   (a) $x^2 + 10x + 15 = 0$ (b) $x^2 - 7x + 9 = 0$
   (c) $7x^2 - 4x - 4 = 0$ (d) $6x^2 + 4x = 3$
   (e) $10x^2 = 2x + 5$ (f) $5x + 1 - 2x^2 = 0$

8. Solve the equations:

(a) $3x^2 + 5x = 2$

(b) $6x^3 - 5x^2 + x = 0$

(c) $x^4 - 5x^2 + 4 = 0$

(d) $x - 3\sqrt{x} + 2 = 0$

(e) $(x + 3)^2 = (2x + 1)^2$

(f) $(x + 13)^2 - 3(x + 13) - 10 = 0$

9. Solve the equations:

(a) $\dfrac{24}{x+1} + \dfrac{9}{x-1} = 5$

(b) $\dfrac{1}{x+1} + \dfrac{1}{x-1} = \dfrac{4}{5}$

(c) $\dfrac{x}{x-2} - \dfrac{x+1}{x+4} = 1$

(d) $\dfrac{x}{x-1} + 1 = \dfrac{x-2}{2x}$

(e) $\dfrac{1}{x^2} - \dfrac{6}{x} + 8 = 0$

(f) $\dfrac{(x-2)(2x+3)}{(x-3)(3x+1)} = 1$

10. For a certain loaded beam the bending moment M is given by

$$M = \frac{(3 + x)^2}{4} - \frac{25}{6}x$$

where x is the distance from one of the supports. Calculate the values of x for which $M = 0$.

Given that the maximum bending moment occurs at $x = \dfrac{16}{3}$, show how to use this fact to check your solution. (N.C.T.E.C.)

11. Plot the graph of $y = x^2 - 3x + 1$ taking values of x from 0 to 4. Use the graph to solve the equation $2x^2 - 6x + 1 = 0$. By superimposing a suitable straight line graph, solve the equation $x^2 - 4x + 2 = 0$..

12. Plot the graph of $y = x^2$ for values of x from -4 to $+4$ and use it to solve the following equations:

(a) $x^2 - x - 1 = 0$

(b) $x^2 = 2(1 - x)$

13. Draw the graph of $y^2 = 4x$ i.e. $y = \pm 2\sqrt{x}$, for values of x between 0 and 9 and thus verify that this is a parabola symmetrical about the x-axis. By adding a suitable straight line, use the graph to solve the equation $x - 2\sqrt{x} - 1 = 0$.

14. Plot the graph of $y = 4x - x^2 + 2$ for x values -2 to $+5$ and use it to solve the equations $4x - x^2 + 2 = 0$ and $4x = x^2 + 1$. For what value of c would the equation $4x - x^2 + c = 0$ have equal roots?

15. Plot the graph of $y = x^2$ for values of x between -5 and $+5$ and use it to find approximately the roots of the equations:

(a) $x^2 + x - 3 = 0$

(b) $x^2 - 3x - 2 = 0$ (W.J.E.C.)

16. Solve graphically:

(a) $x^2 - 4x = 4$

(b) $x^2 - 4x = 8$

[Take values of x from -2 to $+6$] (W.J.E.C.)

Simultaneous equations

Algebraic methods of solving simultaneous linear equations have been outlined in chapter 1, and the same basic methods of elimination and substitution may also be used with equations of higher degree.

Simultaneous equations one of which is a second degree equation

Usually the simplest method is to obtain an expression for one variable from the linear equation and substitute this in the quadratic which will then contain only a single variable and so can be solved by factors or by formula.

Example 14 Solve the simultaneous equations:

$$x^2 - 3xy - 2y^2 = 8 \quad \ldots \quad (1)$$

$$x + y = 1 \quad \ldots \quad (2)$$

From equation (2) we have

$$y = 1 - x$$

Substituting this in equation (1)

$$x^2 - 3x(1 - x) - 2(1 - x)^2 = 8$$

$$x^2 - 3x + 3x^2 - 2(1 - 2x + x^2) = 8$$

$$2x^2 + x - 10 = 0$$

$$(x - 2)(2x + 5) = 0$$

$$\therefore \quad x = 2 \text{ or } -2\tfrac{1}{2}$$

Substitute in equation (2)

$$y = -1 \text{ or } 3\tfrac{1}{2}$$

Example 15 Solve the simultaneous equations:

$$4x^2 + y^2 = 30 \quad \ldots \quad (1)$$

$$x + y = 5 \quad \ldots \quad (2)$$

From equation (2) we have

$$y = 5 - x$$

Substituting this in equation (1),

$$4x^2 + (5 - x)^2 = 30$$

$$5x^2 - 10x - 5 = 0$$

$$x^2 - 2x - 1 = 0$$

$$x = \frac{+2 \pm \sqrt{4 + 4}}{2}$$

$$\therefore \quad x = 1 + \sqrt{2} \text{ or } 1 - \sqrt{2}$$

Substituting in equation (2)

$$y = 4 - \sqrt{2} \text{ or } 4 + \sqrt{2}$$

The graphs of these two equations are shown in fig. 8. The graph of $4x^2 + y^2 = 30$ is seen to be an ellipse, and this is intersected by the straight line $x + y = 5$ at the points (2.41, 2.59) and (− 0.41, 5.41) which correspond to the solutions obtained algebraically.

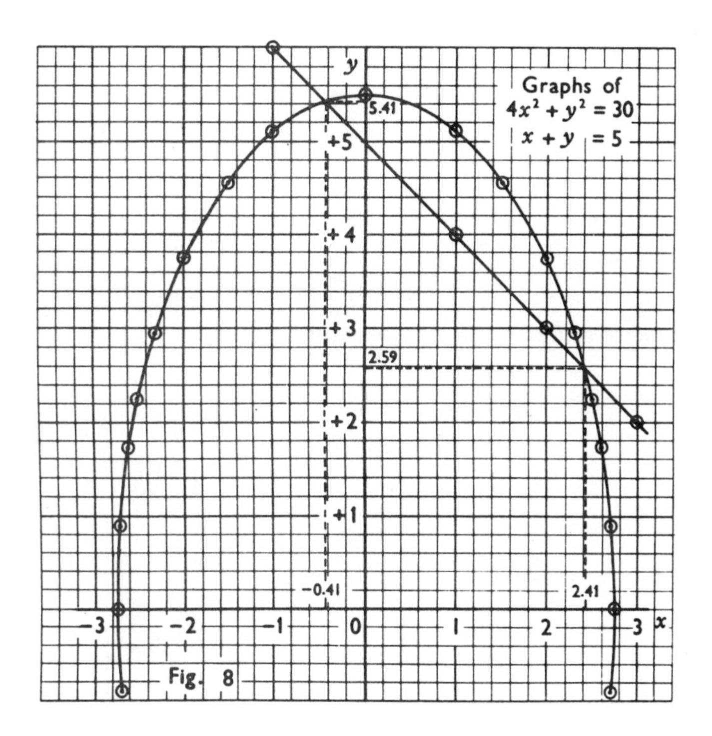

Simultaneous equations both of the second degree

In this case substitution of the usual type would give an equation of the fourth degree and would not normally provide any simplification.

One method which is often very helpful is to try to factorise either one of the original equations or a derived equation so as to split it into two linear equations each of which may be combined with an original equation by the method outlined above. Graphical methods may also be used.

Example 16 Solve:

$$x^2 - xy + y^2 = 19 \qquad (1)$$

$$xy = 15 \qquad (2)$$

Equation (1) can be made into a perfect square if either equation (2) be subtracted from it, or three times equation (2) be added to it.

Subtracting $x^2 - 2xy + y^2 = 4$

$$(x - y)^2 = 4$$

Whence either $x - y = +2$ or $x - y = -2$

$$x = y + 2 \quad \text{or} \quad x = y - 2$$

Substituting in equation (2):

$$y(y + 2) = 15 \qquad y(y - 2) = 15$$

$$y^2 + 2y - 15 = 0 \qquad y^2 - 2y - 15 = 0$$

$$(y - 3)(y + 5) = 0 \qquad (y - 5)(y + 3) = 0$$

$$\therefore \quad y = 3 \text{ or } y = -5 \quad \text{or} \quad y = +5 \text{ or } y = -3$$

Substitute in equation (2)

$$x = 5 \text{ or } x = -3 \text{ or } x = +3 \text{ or } x = -5$$

In this case the solutions may be summarised as:

$$\left.\begin{array}{l} y = \pm 3 \\ x = \pm 5 \end{array}\right\} \quad \text{or} \quad \left.\begin{array}{l} y = \pm 5 \\ x = \pm 3 \end{array}\right\}$$

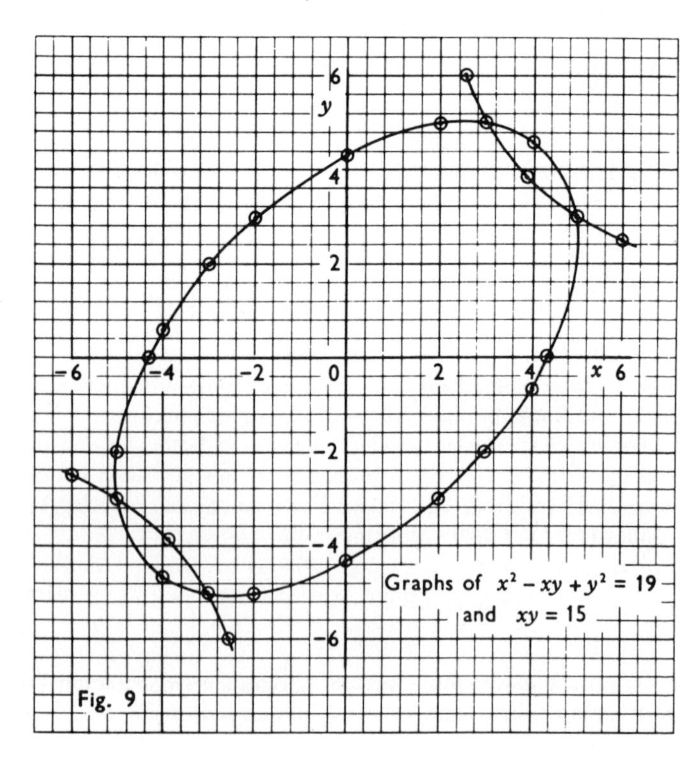

Fig. 9

The graphs of these two equations are shown in fig. 9. The graph of $x^2 - xy + y^2 = 19$ is seen to be an ellipse, and this is intersected by the rectangular hyperbola $xy = 15$ at the four points (5, 3); (3, 5); (– 3, – 5) and (– 5, – 3) which agree with the answers obtained algebraically.

Example 17 Solve:

$$x^2 + xy - 3y^2 = 3 \qquad \qquad (1)$$

$$x^2 - 2xy + 3y^2 = 9 \qquad \qquad (2)$$

This is a case where there are no terms of the first degree in either equation. The method is to cross multiply to eliminate the numerical terms.

Cross multiplying:

$$9x^2 + 9xy - 27y^2 = 3x^2 - 6xy + 9y^2$$
$$6x^2 + 15xy - 36y^2 = 0$$
$$2x^2 + 5xy - 12y^2 = 0$$
$$(2x - 3y)(x + 4y) = 0$$

Hence $\qquad 2x = 3y \quad \text{or} \quad x = -4y$

Substituting in equation (1):

$$\left(\frac{3}{2}y\right)^2 + y\left(\frac{3}{2}y\right) - 3y^2 = 3$$
$$y^2(2\tfrac{1}{4} + 1\tfrac{1}{2} - 3) = 3$$
$$\tfrac{3}{4}y^2 = 3$$
$$y^2 = 4$$
$$\left.\begin{aligned} y &= \pm 2 \\ \text{but } x = \frac{3}{2}y, \text{ therefore } x &= \pm 3 \end{aligned}\right\}$$

$$(-4y)^2 + y(-4y) - 3y^2 = 3$$
$$y^2(16 - 4 - 3) = 3$$
$$9y^2 = 3$$
$$y^2 = \tfrac{1}{3}$$
$$\left.\begin{aligned} y &= \pm \frac{1}{\sqrt{3}} \\ \text{but } x = -4y, \text{ therefore } x &= \pm \frac{4}{\sqrt{3}} \end{aligned}\right\}$$

Solutions of cubics

The cubic equation $x^3 - 1.9x^2 - 3.1x + 4 = 0$ may be solved graphically by various methods. Figure 10 shows the graph of $y = x^3 - 1.9x^2 - 3.1x + 4$ which crosses the axis where $x = -1.6$, 1.0, and 2.5. This method of solving the equation graphically is quite satisfactory provided that neither of the turning points are close to the x-axis.

An alternative quick method is to split the cubic, taking off the term in x and the numerical term to give a straight line graph to intersect the remaining simpler cubic curve. Figure 11 shows the graphs of $y = 3.1x - 4$ and

$y = x^3 - 1.9x^2$ which intersect at the x values of -1.6, 1.0, and 2.5. At the points of intersection,

$$x^3 - 1.9x^2 = 3.1x - 4$$

i.e.

$$x^3 - 1.9x^2 - 3.1x + 4 = 0$$

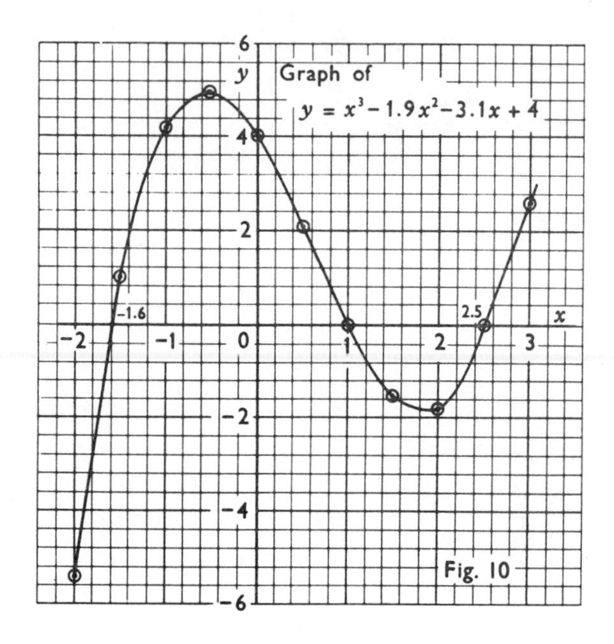

Fig. 10

There are other ways of splitting the cubic and finding a pair of curves to plot, but these two methods are the simplest. Whilst the second method may seem complicated initially, it should be noted that the cubic curve always has a turning point at the origin. This is because any equation of the form $y = ax^3 + bx^2$ can be written as $y = x^2(ax + b)$, and in this form it is easy to see that when the value of y is zero, we get the three solutions of the cubic equation $0 = x^2(ax + b)$ to be $x = 0, 0, -b/a$. Since $y = 0$ when $x = 0$, this curve goes through the origin. Since $x = 0$ twice, the curve touches the x-axis at this point but does not cross it (unless $b = 0$ also). Hence we see that dividing our original cubic in this way gives a simple straight line graph intersecting an easy cubic of known shape.

Closer approximations to roots

The graphical method is to plot to an enlarged scale the portion of the graph in the immediate neighbourhood of the approximate solution. In the previous example, a rough graph by any method would reveal the existence of a solution at approximately $2\frac{1}{2}$, if then points for x values 2.3, 2.4, 2.5, 2.6 and 2.7 were plotted on a much larger scale, the value of the root could be obtained correct to three, if not four, significant figures.

Once one root of a cubic has been obtained to the required degree of accuracy, then this solution may be removed from the equation by long

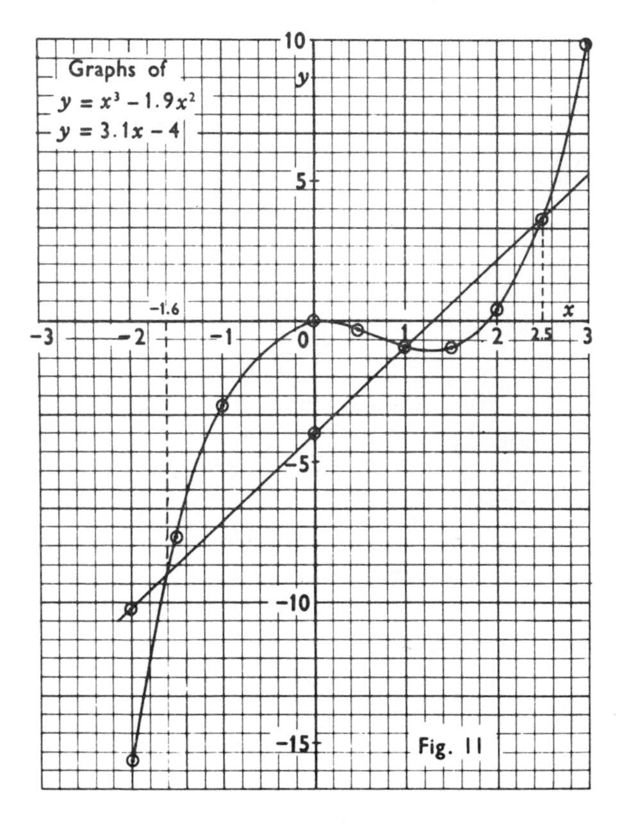

Fig. 11

division leaving a quadratic which may be solved by using the formula. It follows that if the remaining quadratic has no real roots, then the cubic has only one solution.

Example 18 For the equation $x^3 - 1.0x^2 - 3.1x + 4 = 0$ one solution is found to be 2.500. Find the remaining solutions.

If $x = 2.5$ is a solution, then $x - 2.5$ is a factor of the expression $x^3 - 1.9x^2 - 3.1x + 4$. Dividing out:

$$
\begin{array}{r|l}
 & x^2 + 0.6x - 1.6 \\
\hline
x - 2.5 & x^3 - 1.9x^2 - 3.1x + 4 \\
 & \underline{x^3 - 2.5x^2} \\
 & \quad + 0.6x^2 - 3.1x \\
 & \quad\ \ \underline{0.6x^2 - 1.5x} \\
 & \qquad\qquad - 1.6x + 4 \\
 & \qquad\qquad \underline{- 1.6x + 4} \\
 & \qquad\qquad\quad \cdot \quad \cdot \quad \cdot
\end{array}
$$

Thus $x^3 - 1.9x^2 - 3.1x + 4 = (x - 2.5)(x^2 + 0.6x - 1.6) = 0$.

Now using the formula:

$$x = \frac{-b \pm \sqrt{b^2 - 4ac}}{2a} \quad \text{for the quadratic,}$$

$$x = \frac{-0.6 \pm \sqrt{0.36 + 6.4}}{2}$$

$$x = \frac{-0.6 \pm 2.6}{2}$$

$$x = -1.6 \text{ or } +1$$

which are the required solutions.

The algebraic method

A root of an equation of any degree can be found entirely algebraically by the method of successive approximations.

Example 19 Find the smallest positive root of the equation:

$x^3 - 5.25x^2 - 11x + 20 = 0$

Putting $\qquad y = x^3 - 5.25x^2 - 11x + 20$

when $x = 0$, $\qquad y = 0$

when $x = 1$, $\qquad y = 1 - 5.25 - 11 + 20 = 4.75$

when $x = 2$, $\qquad y = 8 - 21 - 22 + 20 = -15$

Since the value of y is positive when $x = 1$ and negative when $x = 2$ the graph, if plotted, would cross the x-axis somewhere in between. As the value of y is numerically smaller for $x = 1$ than for $x = 2$, it is reasonable to assume that the solution is closer to $x = 1$ than to $x = 2$, and we therefore take $x = 1$ as our first approximation.

For a closer approximation, let $x = 1 + h$, where h is a small fraction less than $\frac{1}{2}$. Putting this value in our equation, we have:

$$(1 + h)^3 - 5.25(1 + h)^2 - 11(1 + h) + 20 = 0$$

i.e. $$1 + 3h + 3h^2 + h^3 - 5.25(1 + 2h + h^2) - 11(1 + h) + 20 = 0$$

Now since h is a small fraction, h^2 and h^3 will be smaller still and may be disregarded.

$$1 + 3h - 5.25 - 10.5h - 11 - 11h + 20 = 0$$
$$18.5h = 4.75$$
$$h = 0.257$$

Therefore for a closer value we take as our second approximation

$$x = 1 + 0.26$$

and proceed as before.

Putting in $x = 1.26 + h$, we have

$$(1.26 + h)^3 - 5.25(1.26 + h)^2 - 11(1.26 + h) + 20 = 0$$

disregarding terms in h^2 and h^3 as before:

$$1.26^3 + 3.78h - 6.615 - 10.5h - 13.86 - 11h + 20 = 0$$
$$17.72h = -1.525$$
$$h = -0.0086$$

Thus our third approximation is

$$x = 1.26 - 0.0086 = 1.2514$$

and the solution is

$$x = 1.25$$

correct to 3 significant figures.

Alternatively, we could proceed in a somewhat different way as follows: First of all put $X + 1$ for x in our equation. (This is equivalent to transferring the origin to the point $x = 1$, $y = 0$.)

Thus $\qquad (X + 1)^3 - 5.25(X + 1)^2 - 11(X + 1) + 20 = 0$

i.e. $X^3 + 3X^2 + 3X + 1 - 5.25X^2 - 10.5X - 5.25 - 11X - 11 + 20 = 0$

which simplifies to

$$X^3 - 2.25X^2 - 18.5X + 4.75 = 0$$

Put this as

$$18.5X = 4.75 - 2.25X^2 + X^3$$

and multiply by 4 to clear the fractions

$$74X = 19 - 9X^2 + 4X^3$$

By transferring the origin to where $x = 1$, we are now only a small fraction away from the required root and for a first approximation may neglect X^2 and X^3. Therefore the first approximation is given by:

$$74X = 19$$
$$X = 0.257$$

For our second approximation we include the term in X^2:

$$74X = 19 - 9X^2$$

Putting in the first approximation in the R.H.S. only:

$$74X = 19 - 9(0.257^2)$$
$$74X = 19 - 0.5945$$
$$X = 0.248\,64$$

giving us a second approximation of:

$$X = 0.249$$

For our third approximation we include also the term in X^3:

$$74X = 19 - 9X^2 + 4X^3$$
$$74X = 19 - 9(0.249)^2 + 4(0.249)^3$$
$$74X = 18.502$$
$$X = 0.250\,03$$

but $x = X + 1$,

$$\therefore \quad x = 1.2500$$

correct to 4 decimal places.

This method is particularly useful when the root is close to the origin.

Any of these methods may be used to approximate to the roots of equations of higher degree.

The tedious repetition involved in successive approximation can be reduced considerably if a calculating machine is available. The method is given in the next chapter.

EXERCISE 3(b)

1. (a) Solve the simultaneous equations:

$$\frac{2}{x} - \frac{1}{y} = \frac{1}{3}$$
$$\frac{5}{x} - \frac{3}{y} = \frac{1}{2}$$

(b) The sum of two numbers is 28. The sum of their squares is 400. Prove that the difference of their squares is 112.

2. A builder complained that a delivery of bricks was short by an equal number of blue and rustic bricks. The order had been for a total of 7875 bricks. If 20% of the blue bricks and 1% of the rustic were missing, calculate how many blue bricks were ordered. (U.L.C.I.)

3. Solve:

$$a + 2b = 4$$
$$a^2 - 2b^2 = 32$$

(K.C.E.A.B.)

4. Solve the simultaneous equations:

$$(x + 1)(2y + 1) = 4xy$$
$$x - \tfrac{1}{3}y = 1$$

(U.E.I.)

5. Solve the simultaneous equations:

$$x^2+xy+y^2 = 3$$
$$x^2+2xy+3y^2 = 3$$

6. (a) Solve the equations:

$$y = 2(x-2)$$
$$y^2 - x^2 = 2x - 4y - 7$$

(b) The values of x and y given in the table satisfy the equation:

$$y = ax^2 + bx + c$$

Calculate the values of the constants a, b and c.

x	3	4	-1
y	0	11	-4

Hence obtain the value of y when $x = 1$. (W.J.E.C.)

7. Solve the simultaneous equations:

$$3x^2 - 3y^2 = 8xy$$
$$x + 2y - 10 = 0$$

(W.J.E.C.)

8. Plot the graph of $y = x^3 - 2x^2 - 5x + 6$ and hence solve the equation $x^3 - 2x^2 - 5x + 6 = 0$. (E.M.E.U.)

9. Plot, on the same axes, the graphs $y = \frac{1}{3}x^3$ and $y = 2x + 1$ for values of x between -3 and $+3$. Use the graphs to find approximate values of the roots of the equation, $x^3 - 6x - 3 = 0$. (U.L.C.I.)

10. Solve graphically the equation $x^3 - 2x^2 - 11x + 12 = 0$. (K.C.E.A.B.)

11. The equation $y = x^3 + 2x^2 - 4x + 7$ has one real root. Obtain values of y for values of x from -5 to $+3$ and plot the curve. By a graphical process, obtain the value of the real root correct to 3 significant figures. (N.C.T.E.C.)

12. In calculating the velocity of flow of a liquid along a channel, using Bernoulli's Theorem, the following equation giving the velocity V in m/sec arises:

$$V^3 - 360V + 2400 = 0$$

Show that there is a root of this equation between $V = 8$ and $V = 9$, and find this root to two decimal places. (D.D.C.T.)

REVISION EXERCISE 3

1. Solve the quadratic equation: $x(x + 6) = -6$ (U.L.C.I.)

2. Solve, correct to one place of decimals: $x^2 - 5x = 10$ (U.L.C.I.)

3. (*a*) Solve the following equations:

(i) $x^2 - 1 = 0$ (ii) $6x^2 + 7x - 20 = 0$

(b) Express x in terms of p and q if $x(x + 2p) = q$ (U.L.C.I.)

4. Solve, correct to one place of decimals, the equation:

$$\frac{3x^2}{4} - x = \frac{1}{3}$$

(U.L.C.I.)

5. Solve by factorisation:
   (a) $x^2 - 25x - 54 = 0$ (b) $6x^2 - x - 15 = 0$ (N.C.T.E.C.)

6. Solve the equation:

$$\frac{1}{x^2 + 2x - 3} + \frac{2}{x - 1} = 2.2$$ (N.C.T.E.C.)

7. Solve:
   (a) $x(x + 7) = 7(x + 28)$
   (b) $x^2 - 28x + 187 = 0$ (K.C.E.A.B.)

8. Solve the equation: $(x - 1)^2 + 3(x - 1) - 1 = 0$ (U.L.C.I.)

9. Solve the equations:
   (a) $3x^2 + 4x - 7 = 0$ (b) $3x^2 - 4x - 8 = 0$ (K.C.E.A.B.)

10. Solve the equation $2x^2 = 3 - 7x$, (i) correct to two places of decimals by using the formula, *or* (ii) as accurately as possible by drawing a graph. (U.L.C.I.)

11. (a) Solve the equation:

$$(2x - 1)(2x + 3) + (2x + 1)(2x + 3) = 2(2x + 1)(2x - 1)$$

   (b) Factorise:
   (i) $(1 - x)^2 - xy + y$
   (ii) $(ax + by)^2 - (ay + bx)^2$ (U.E.I.)

12. In a simply supported beam with a uniform load the bending moment M at a distance x from one end is given by:

$$M = \frac{Wx}{2l}(l - x)$$

where W is the load and l the length of the beam. Calculate the values of x if $l = 16$, $M = 3$, and $W = 2$. (U.E.I.)

13. A formula used in finding the strength of a concrete beam is

$$bn^2 + 2am(n - c) = 0$$

   (a) Express n in terms of the other letters.
   (b) Find n if $b = 4.5$, $a = 1.7$, $c = 8$, and $m = 12.5$. (K.C.E.A.B.)

14. (a) Solve, correct to two places of decimals:

$$6x^2 - 11x - 9 = 0$$

   (b) Prove $2x^2 - 3x - 3$ is a factor of:

$$8x^4 - 12x^3 - 30x^2 + 27x + 27$$

   and find the remaining factors. (U.L.C.I.)

15. (a) Solve, correct to two places of decimals, the equation:

$$3x^2 - 4x - 1 = 0$$

   (b) A rectangular lawn of length 5 m and breadth 3 m has its length increased by x m, and its breadth by $2x$ m. The area of the lawn is thereby doubled. Form an equation in x and solve it. (U.L.C.I.)

16. Plot the graph of the equation $y = 2x^2 - 7x + 1$ between the values $x = -1$ and $x = +4$. From your graph, read the roots of the equation:

$$2x^2 - 7x + 3 = 0$$

Check your solution algebraically. (K.C.E.A.B.)

17. Draw the graph of $y = 8x^2 - 10x - 55$ from $x = -4$ to $x = +5$. From your graph read off the solutions of the equation $8x^2 - 10x = 55$. Check your results algebraically. (K.C.E.A.B.)

18. Plot a graph of the curve $y = x^2$ from $x = -4$ to $x = +4$. By drawing approximate straight lines to cut this curve solve the equations:
(a) $x^2 - x - 6 = 0$ (b) $3x^2 + 2x - 18 = 0$ (U.L.C.I.)

19. With the same axes draw the curve $y = 3x^2 - 2$ and the straight line

$$y = \frac{10(5x + 21)}{11}$$

using suitable values of x between -4 and $+4$. Use your graph to solve the equation $33x^2 = 50x + 232$. (U.E.I.)

20. Draw the graph of $y = x^2 - x - 6$ and state the values of x for which $y = 0$. (N.C.T.E.C.)

21. (a) From the equations given, find a, b and c, the angles at a station in a survey.
(i) $4a + 2b + c = 287° 7'$
(ii) $2a + 7b + 3c = 496° 39'$
(iii) $a + 4b + 4c = 402° 41'$
(b) Find the values of x and y, given $x^2 + y^2 = 58$ and $x - y = 4$.
(N.C.T.E.C.)

22. (a) The combined heights of two unequal cubes are 11 metres and their combined volumes equal 407 m^3. Calculate the dimensions of each cube.
(b) Solve the following simultaneous equations:

$$a + b + c = 11$$
$$3a - b + 2c = 9$$
$$a + 3b - c = 13$$

(K.C.E.A.B.)

23. (a) Find a, b and c from the linear simultaneous equations:

$$a - 2b + 2c = 1$$
$$2a + b - 8c = 6$$
$$a - 3b + 6c = 1$$

(b) Solve the simultaneous equations:

$$x - 3y = 2$$
$$x^2 - 12y = 13$$

(W.J.E.C.)

24. Solve the following pairs of simultaneous equations:
(a) $(x + 1)(2y + 1) = 4xy$, $x - \frac{1}{3}y = 1$
(b) $x^2 - 3xy + 2y^2 = 40$, $2x - 3y = 13$
(c) $x^2 - 4y^2 = 8$, $2(x + y) = 7$
(U.E.I.)

25. In the equation:

$$\frac{V}{\pi} = 5h^2 - \frac{h^3}{3}$$

V m³ is the volume of liquid contained in a spherical vessel of 10 m diameter, when the depth of liquid at the centre is h m. Plot a graph showing the connection between V/π and h, for values of h between 0 and 10 m. Use the graph to find the value of h when $V = 440$ m³. (U.L.C.I.)

26. Find, by graphical means, the roots of the equation $9x^2 - 2x^3 = x + 12$, given that they lie within the range $x = -2$ and $x = +5$. (U.L.C.I.)

27. Draw the graph of $y = x^3 + x^2 - 8x - 7$ for values of x from -3 to $+3$ and hence solve the equation:

$$x^3 + x^2 - 8x - 7 = 0$$

By drawing a portion of the graph to a larger scale, find the positive root of this equation correct to two decimal places. (W.J.E.C.)

28. In the design of a tubular steel column subjected to a lateral thrust, the internal diameter was given by the equation:

$$\frac{(d + 1)^4 - d^4}{d + 1} = 50$$

Expand the equation and simplify to a cubic. Replace d by $(x + h)$. Equate the coefficient of x^2 to zero and show that the numerical value of h so obtained leads to the equation $4x^3 - 49x - 25 = 0$.

Solve this equation graphically to obtain the value of the positive root correct to the first decimal place. (N.C.T.E.C.)

29. The deflection D of a girder of length x, of cross-sectional area A, with a load W concentrated at its middle section is given by the equation:

$$D = \frac{Wx}{8AE}\left[\frac{x^2}{6k^2} + x\right]$$

where $E = 3 \times 10^7$.

Show that the length of the girder for which $A = 4$, $k^2 = 24$, $W = 4000$, and $D = \frac{1}{48}$ is a root of the equation $x^3 + 720x - 720\,000 = 0$.

Show that this root lies between 86 and 87 and find the root to one decimal place. (D.D.C.T.)

30. Plot the graph $y = 1 - 2x - 3x^2$ for values of x between ± 3.

(a) Use the graph to solve the equations

(i) $1 - 2x - 3x^2 = 0$ (ii) $3x^2 + 2x - 16 = 0$

(b) From the graph determine the maximum value of

$$y = 1 - 2x - 3x^2$$

(N.C.T.E.C.)

31. Plot the graph of $y = x^3 + x^2 - 8x - 5$ for values of x from -4 to $+3$ and hence solve the equation $x^3 + x^2 - 8x - 5 = 0$.

By drawing a portion of the graph to a larger scale, find the positive root of the equation correct to two decimal places. (U.E.I.)

4 Computation

ATTEMPTS to speed up the process of calculation have resulted in the invention of various aids which are still in use today. The **abacus** was invented before 500 BC but is still in common use in Asia and Eastern Europe. Napier invented **logarithms** in 1614 and logarithmic tables and the slide rule soon followed. In more recent times there has been a development from the earlier types of **mechanical** calculators, through **electric** machines, on to **electronic** calculators and, finally, to **computers.**

All calculators are limited in what they can do. The basic operations are addition and subtraction and all other operations usually have to be built up from these. Thus multiplication can be done by repeated addition, and division by repeated subtraction. More complicated processes have to be broken down into simple steps of addition or subtraction for the calculator or computer and this process of planning the sequence of required operations is known as **programming.**

This first exercise is included for those who have available some form of simple electric or electronic calculator. No answers are provided to the questions in this exercise because it is intended that each calculation should be attempted on a machine and checked by doing the same calculation with the use of slide rule, logarithmic tables, or tables of squares or reciprocals as appropriate.

EXERCISE 4(a)

1. $125 + 873 - 257$
2. $23\,500 + 60\,700 + 32\,800$
3. $1.45 + 7.98 + 3.54$
4. $8.96 - 0.73 + 1.02$
5. 259×107
6. $20\,480 \div 25.6$
7. $4.58 \times 6.13 \div 9.16$
8. $(2.065)^2$
9. $(8.62)^2 - (7.28)^2$
10. 0.065×0.008
11. $\frac{3}{4}(72 \times 43)$
12. $\frac{2}{3}(8.4 \times 7.3)$
13. $\frac{2}{3}(678 + 326) - \frac{1}{3}(215 + 173)$
14. $\frac{1}{2}(6.842)^2$
15. $\frac{7}{8}(4.096)^2$
16. 1/11 to 4 decimal places
17. 1/0.008

18. 5% of 640
19. 12% of 156.72
20. $\frac{21.4 \times 4.8}{2.88} - \frac{12.5 \times 6.4}{3.84}$
21. Find the value of πr^2 when $\pi = 3.14159$ and $r = 5.12$
22. Find the value of $\frac{1}{3}\pi r^2$ when $\pi = 3.1416$ and $r = 2.06$
23. Find the value of $x^2 + 5x + 3$ when $x = 1.22$
24. Find the value of $x^2 - 0.42x + 7.86$ when $x = 0.928$
25. Evaluate $\frac{x^3 + 4x - 5.2}{(x - 1)^2}$ when $x = 2.08$

Evaluation of polynomials

In the last chapter, we showed that the equation

$$x^3 - 5.25x^2 - 11x + 20 = 0$$

had a root between $x = 1$ and $x = 2$. An expression such as

$$x^3 - 5.25x^2 - 11x + 20,$$

which contains only positive integral powers of the variable, is called a **polynomial**. The most effective way of using a calculating machine to evaluate a polynomial involves a process known as **nesting**. For this purpose we factorise the polynomial as follows:

$$x^3 - 5.25x^2 - 11x + 20 = x[x(x - 5.25) - 11] + 20$$

This shows us the following sequence of steps:

(a) put in x
(b) subtract 5.25
(c) multiply by x
(d) subtract 11
(e) multiply by x
(f) add 20

For $x = 1$, the nesting program gives the following result:

(a) put in x	1.00
(b) subtract 5.25	−4.25
(c) multiply by x	−4.25
(d) subtract 11	−15.25
(e) multiply by x	−15.25
(f) add 20	+4.75

For $x = 2$, we get the following result:

(a) put in x	2.00
(b) subtract 5.25	−3.25
(c) multiply by x	−6.50
(d) subtract 11	−17.50
(e) multiply by x	−35.00
(f) add 20	−15.00

To get closer to that value of x which will give a final remainder of zero, and hence the required solution, we must repeat this sequence several times. Such a process, designed to bring us ever closer to the required solution, is called an **iterative** process and is ideal for use with a calculator or computer. The sequence of operations is usually expressed in a diagram called a **flowchart.** A flowchart consists of a series of connected symbols designed to illustrate the sequence of operations. Words, letters, or numbers, may be inserted within the symbol to explain the function of that particular symbol, but all such text must be kept to an absolute minimum. Our nesting program can now be put in the form of a flowchart as shown (fig. 12).

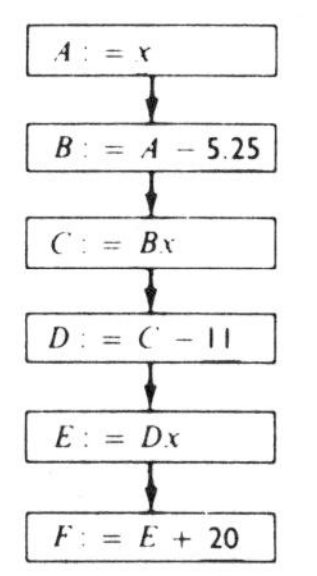

Fig. 12

When we tried the values $x = 1$ and $x = 2$ we found that the value of the polynomial $x^3 - 5.25x^2 - 11x + 20$ changed from $+4.75$ to -15.00 as x changed from 1 to 2. This suggests that the value of x we are looking for should lie between 1.0 and 2.0 and it would be reasonable to try an intermediate value such as $x = 1.5$ which causes our program to yield the following result:

$$\begin{aligned} A &= 1.5 \\ B &= -3.75 \\ C &= -5.625 \\ D &= -16.625 \\ E &= -24.9375 \\ F &= -4.9375 \end{aligned}$$

From this we see that the value of the polynomial has changed from $+4.75$ to -4.94 as x changed from 1.0 to 1.5 and it would therefore be logical to take the mid-value of $x = 1.25$ yielding the following result from our program:

$$\begin{aligned} A &= 1.25 \\ B &= -4.00 \\ C &= -5.00 \\ D &= -16.00 \\ E &= -20.00 \\ F &= 0 \end{aligned}$$

Hence the value $x = 1.25$ is the required solution.

The procedure we have followed may seem rather lengthy at first sight. There are two reasons for this. Firstly, whenever we tried a value, we put in all the intermediate stages of the working, whereas we did not really need to know the values of B, C, D and E. All we need is to put in a value for x at stage A and see whether it gave us $F = 0$. Secondly, after following through our program with one particular value for x, we paused to consider what to choose for our next trial value. The whole process would be faster if the choice of the next value could be determined by an automatic process. What we need is a loop in the program specially designed to select the next value and insert it into the start of the sequence.

Our modified program flowchart now looks like fig. 13. If this program is put into a computer or programmable calculator, it is now only necessary to feed in the upper value of $x = 2.0$ and the computer will run through the

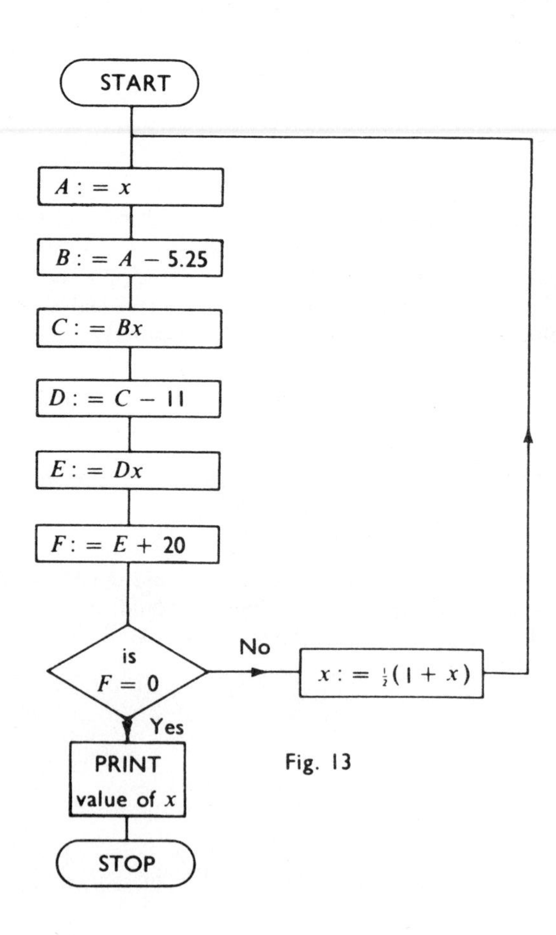

Fig. 13

program automatically as many times as required and in less than a second it should produce the result $x = 1.25$.

Saving computer time

Look closely again at our program. Can you see any point where we could have wasted time with this program on the computer? Our program was designed to run until we obtained a value for F which was *exactly zero*, without any regard for how many times the loop sequence was repeated or how long it took! Whilst in this case we knew that the exact value of $x = 1.25$ would soon be reached, in general it is necessary to ensure that we do not waste valuable computer time finding a value for x accurate to more decimal places than we need. This can be done by arranging the program to run only until a result is achieved within acceptable limits of accuracy.

Square root program

This example has been selected because it illustrates the point that any iterative process can be continued until the successive approximations reach whatever level of accuracy is required. This program will enable us to find the square root of any number and we can specify how many significant figures we would like in the answer. Let us call the number N of which we want to find the square root. Let us specify an accuracy of three decimal places. A program for this could then be written as in fig. 14. Suppose we test our program by using it to find the value of $\sqrt{3}$. If we follow through the various stages, the flowchart (fig. 14) yields the following sequence.

(a) Put in the value $N = 3$.
(b) Choose a first value for x. Since our value for N lies between 1 and 4, we deduce that its square root must lie between 1 and 2 and probably it will be closer to 2. Therefore, let $x = 2$ at this stage.
(c) N/x is now computed as $\frac{3}{2}$, i.e. 1.50.
(d) Is $2.00 > 1.50$? Yes, therefore $D = 2.00 - 1.50 = 0.50$.
(e) Is $0.50 < 0.0001$? No, therefore make $x = \frac{1}{2}(2.00 + 1.50) = 1.750$.

The loop now brings us back for the second run through the sequence. Now

$$B = 1.750 \quad \text{so} \quad C = 3 \div 1.750 = 1.714$$

Then

$$1.750 > 1.714 \quad \text{so} \quad D = 1.750 - 1.714 = 0.036$$

This is still not accurate enough, so the process is repeated using our new value of

$$x = \tfrac{1}{2}(1.750 + 1.714) = 1.732$$

Finally we check through the sequence again until our program reveals that we have a value for $\sqrt{3}$ which is sufficiently accurate and this is the value which is printed out for us as the final solution.

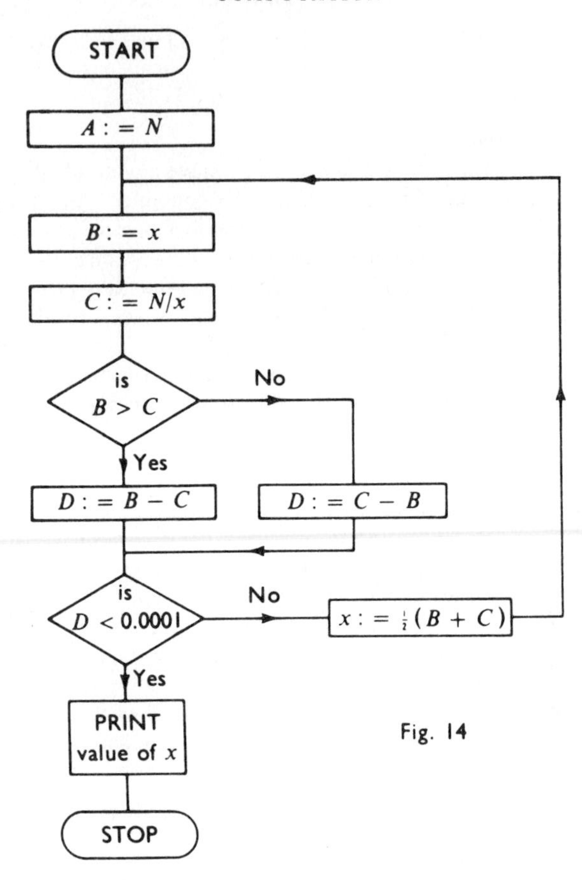

Fig. 14

Flowchart symbols

In the flowchart we have just used, we not only employed the rectangular symbol

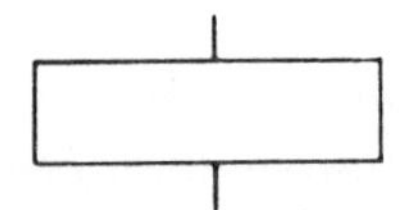

which represents a processing function, but we also introduced other symbol shapes to indicate other processes. Throughout this chapter we shall adopt the British Standard specification for flowchart symbols (BS 4058:1973). These symbols may be enlarged or reduced in size, but it is extremely important that they should always be drawn to the correct proportions. Stencils of the correct shapes are available in various sizes and it is recommended that a standard stencil of flowchart symbols should be used in order that confusion be avoided.

The functions of the other symbol shapes used so far are as follows:

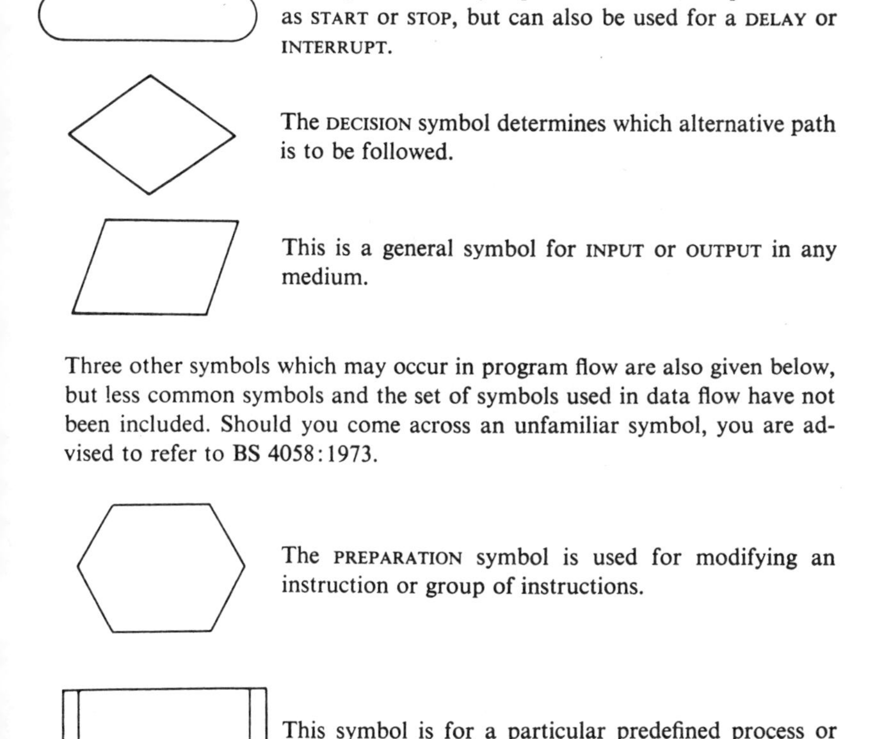

This symbol usually represents a terminal point such as START or STOP, but can also be used for a DELAY or INTERRUPT.

The DECISION symbol determines which alternative path is to be followed.

This is a general symbol for INPUT or OUTPUT in any medium.

Three other symbols which may occur in program flow are also given below, but less common symbols and the set of symbols used in data flow have not been included. Should you come across an unfamiliar symbol, you are advised to refer to BS 4058:1973.

The PREPARATION symbol is used for modifying an instruction or group of instructions.

This symbol is for a particular predefined process or sub-routine.

When it is desirable to break a flow line to avoid crossovers, or when a flowchart is to be continued on another page, this CONNECTOR symbol is used.

Whatever symbols are employed, it is important to keep the text inserted within a symbol as brief as possible. The examples already given show how the instructions are almost restricted to a mathematical shorthand.

Where this is not possible, because additional explanation is necessary or some other descriptive comment is required, an ANNOTATION symbol can be used.

Flow lines

In a program flowchart the symbols are connected by flow lines to indicate the sequence of operations. It is conventional to maintain the direction of flow either horizontally from left to right or vertically from top to bottom. If the flow is not in a standard direction, or there is any doubt, arrows should be used on the flow lines to indicate direction.

When two flow lines have to cross one another, one line is simply drawn across the other at right angles and no junction is implied. Where flow lines are intended to join, a T-junction is used.

Non-mathematical examples

The same principles of programming apply to the sequences of operations and alternative decisions, etc., which occur in a great variety of non-mathematical situations. Having started with some mathematical examples because these are easier to follow through in the various stages, we shall now consider other flowcharts to indicate how the methods can be applied to problems in construction or administration.

Making out a flowchart can be a useful exercise in itself because it can often reveal that some procedures accepted as part of normal routine are capable of modification to improve efficiency, or, in some cases, whole sections of repetitive office record systems can be omitted altogether!

The greatest advantage, however, is when the flowcharting is designed to enable as many as possible of the routine tasks to be taken over by the computer. A computer can check through past records at such great speed, that the expense of computer time can be well worth while when compared with the expense of paying office staff to do the same task by manual sorting. Larger construction firms are now using computers to produce detailed quotations for complicated contracts in a fraction of the time it used to take a team of surveyors. Whenever standard sub-routines can be used, the computer proves its value in such a variety of different roles, yet merely acquiring a computer will not be much help unless its full potentialities are utilised in such a way that computer time is not wasted.

Records

To illustrate the problem of using a computer to search through a set of record cards (or information on tape) consider the problem of looking up a word in a dictionary. The first problem is to ensure that we have the right dictionary, for there are dictionaries in many languages: so, with our computer, we must select the correct set of records (usually by a codename). Assuming we now have the right dictionary, how do we locate the word we require? Should we start at the first page and examine every word listed in order until we finally reach the word we seek? It would be easy to program a computer to do the equivalent of that, but evidently it would be a very inefficient method. It is fairly obvious that we would first look at our required word and see what was its first letter. We could then proceed rapidly past all sections of the dictionary where the words started with letters further back in the alphabet than the letter we were seeking. Repeating this with the second and subsequent letters of our word would rapidly bring us to the relevant entry in the dictionary.

Note that in this example we have a case where we could either have a very simple program which wasted a lot of computer time, or, alternatively, we could construct a more complicated program which would be much more efficient. From the information given, you should now be able to draw and compare the two flowcharts.

Construction programs

In planning any construction process, the first essential is to get all the sectional activities into logical sequence. If only one operation is involved, this may be fairly straightforward. The purpose is to ensure that no subsequent activity is delayed unnecessarily simply because someone forgot the need to order the materials in time! With more complex operations it is apparent that several different activities are in progress simultaneously and the purpose of the programming is to ensure that no men are idle whilst waiting for others to complete the preliminary stages. For example, on a simple house extension job, it would be useless to send along a painter, a plasterer, an electrician and a joiner and expect them all to work in the same room on the same day in an attempt to get the job done quickly!

The methods of network analysis devised to deal with the progressing of more complex building operations is beyond the scope of the present volume, so at this stage we shall confine ourselves merely to the problems of arranging operations in sequence. Even so, some degree of decision-making may well arise, should it be necessary to allow for such contingencies as ill-health, bad weather, non-arrival of materials, etc.

Sequence of activities

Taking an extremely simple example, let us consider the operation of laying a length of drainpipe. Suppose we identify the following stages:

(1) Excavate trench to required depth.
(2) Place drain pipes in position.

(3) Seal all joints.
(4) Backfill trench.

Even in this elementary example we can see that the actual progress could be made in more than one way and one factor which could influence the sequence would be whether the work was to be done by one man or by more than one. What other factors might be involved? Your consideration

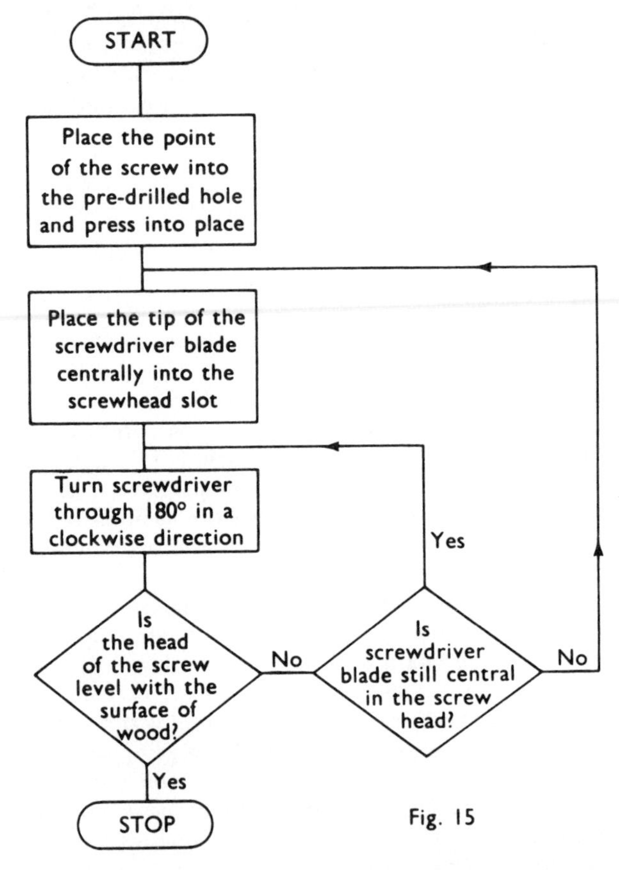

Fig. 15

(or class discussion) might centre around some of the following possibilities: Does it make any difference to the sequence if an excavator is hired for stage (1)? How soon after stage (1) is started can stage (2) begin? If it is to be a one-man job, should he alternate his activities to provide variety? If there are two men on the job, how is it affected by their relative skills and experience? Suppose the trench runs across a main road? What would be the effect of bad weather?

By now you have probably realised that although the operation could be carried out exactly in the sequence we first stipulated, there may be circumstances in which it would be advisable to start one stage before the previous one had been completed.

Flowchart example

Not all activities prove to have as many complications as some of the realistic examples we have just considered. Whilst it is not easy to devise examples which are simple and yet still realistic, let us take one typical example. Can you devise a simple flowchart to illustrate how to screw an ordinary woodscrew into a piece of timber? The flowchart should resemble fig. 15.

EXERCISE 4(b)

1. Expand $x[x\{x(x-2)-3\}-4]$
2. Evaluate $x[x\{x(x+1)-2\}+3]$ when $x = 2.20$
3. Find the root of the equation $x^3 - 1.75x^2 - 6.75x + 11 = 0$ between $x = 2$ and $x = 3$.
4. Find the root of $x^3 - 0.07x^2 - 1.90x + 0.33 = 0$, taking $x = 1.50$ as a first approximation.
5. Use the square root program to find the square roots of each of the following numbers, giving your answers to four significant figures:
   (a) 71.88 (b) 0.8866 (c) 0.054 78 (d) 0.000 007
6. Identify the various activities involved in the construction of an interior wall of brick and plaster and list them in order.
7. List in order the various activities involved in laying a tarmac drive.
8. Use the relation $(n+1)^2 = n^2 + 2n + 1$ to construct a flowchart to find the value of $(n+1)^2$ from the value of n^2 and use it to find the values of 101^2, 102^2, and 103^2.
9. Use a similar technique to that in the previous question to devise a flowchart for calculating $(n-1)^2$ and use it to find 99^2 and 98^2.
10. The following table shows values of n^3, and each subsequent column is derived by differencing adjacent figures in the previous column. Fill in the gaps shown bracketed.

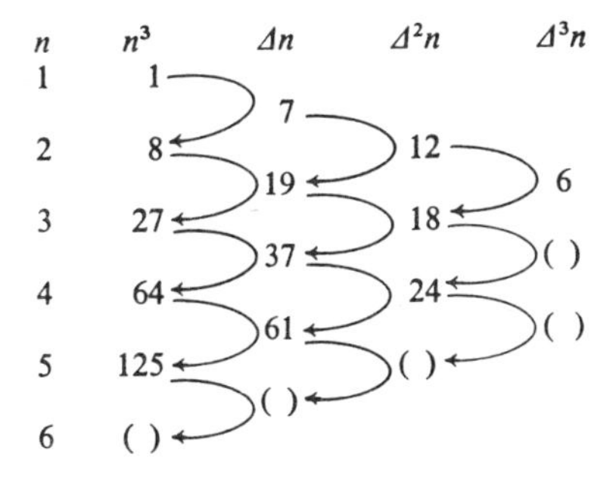

REVISION EXERCISE 4

1. Simplify the following expression as far as possible

$$1 - x[1 - x\{1 - x(1 - x)\}]$$ (U.E.I.)

2. Use the nesting process to evaluate $2x^3 + 3x^2 - 9x + 2$ when $x = 1.4$.
3. Devise a simple flowchart for a program to put two given numbers in order of increasing size. (U.E.I.)
4. Organise the following instructions into flowchart form for the process of precast concrete manufacture (these are *not* in order):
   (1) erect formwork;
   (2) raw materials for concrete-input;
   (3) steel reinforcing rod-input;
   (4) cure concrete;
   (5) pour concrete;
   (6) strip form-work;
   (7) finished product output;
   (8) bend reinforcement;
   (9) mix concrete;
   (10) clean and oil formwork;
   (11) place reinforcement. (N.C.T.E.C.)
5. Given that $120^2 = 14\,400$, find 119^2 and 121^2.
6. Draw a flowchart for one of the following building operations:
   (a) Using a concrete mixer to produce a barrow-load of concrete suitable for a footpath.
   (b) Erecting scaffolding in preparation for re-pointing the brickwork on a shaky chimneystack.
7. Using as successive stages in a program flowchart

   $A: = Nx$
   $B: = 2 - A$
   $C: = Bx$

   complete a program for calculating the reciprocal of any number N to four significant figures. Use your program to evaluate $\frac{1}{7}$ starting with $x = 0.100$ as a first approximation.
8. (a) Explain any THREE of the following flowchart symbols:
   (i) ◇ (ii) ▱
   (iii) ⬭ (iv) ⬡
   (b) Draw a flowchart, with at least one branch, for a particular simple building operation of your own choice. (U.E.I.)

5 Basic Trigonometry

Fundamental facts

Consider carefully the right-angled triangle ABC shown in fig. 16. Although the letters A, B and C are put there to label the points where the sides of the triangle intersect (called *vertices*), the angles of the triangle are also often referred to by letters. Thus angle A, or $\angle A$, may be used to denote the

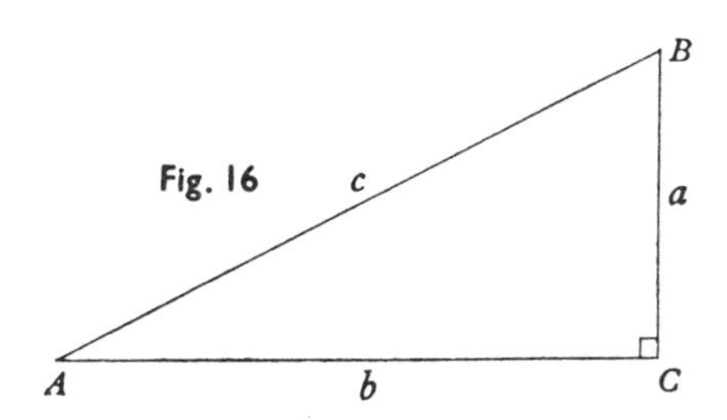

Fig. 16

angle at vertex A. If this is likely to cause confusion, e.g. if there should be more than one angle at a particular point, then it is better either to refer to it as angle BAC, i.e. the angle between BA and AC, or else to place a small letter such as x, y, z, α, β, θ or ϕ, in the angle so that the particular angle referred to may be easily distinguished.

In any right-angled triangle the side opposite to the right-angle is called the **hypotenuse**, the other two sides being called the **opposite** and the **adjacent** in accordance with their position relative to the angle under consideration. Thus in fig. 16, relative to angle A the side BC is opposite and AC is adjacent, whereas it is the reverse relative to angle B. It is usual to denote the side opposite angle A as a, the side opposite angle B as b, etc.

The three most important trigonometrical ratios are defined as follows:

(1) The **sine** of angle A: $\sin A = \dfrac{\text{opposite}}{\text{hypotenuse}}$

(2) The **cosine** of angle A: $\cos A = \dfrac{\text{adjacent}}{\text{hypotenuse}}$

(3) The **tangent** of angle A: $\tan A = \dfrac{\text{opposite}}{\text{adjacent}}$

The reciprocals of these are also very useful, and these are as follows:

(4) The **cosecant** of angle A: $\text{cosec } A = \frac{1}{\sin A}$

(5) The **secant** of angle A: $\sec A = \frac{1}{\cos A}$

(6) The **cotangent** of angle A: $\cot A = \frac{1}{\tan A}$

Let us now state these ratios for angles A and B of fig. 16 in terms of the sides of the triangle.

$$\sin A = \frac{BC}{AB} = \frac{a}{c} \qquad \sin B = \frac{AC}{AB} = \frac{b}{c}$$

$$\cos A = \frac{AC}{AB} = \frac{b}{c} \qquad \cos B = \frac{BC}{AB} = \frac{a}{c}$$

$$\tan A = \frac{BC}{AC} = \frac{a}{b} \qquad \tan B = \frac{AC}{BC} = \frac{b}{a}$$

$$\text{cosec } A = \frac{AB}{BC} = \frac{c}{a} \qquad \text{cosec } B = \frac{AB}{AC} = \frac{c}{b}$$

$$\sec A = \frac{AB}{AC} = \frac{c}{b} \qquad \sec B = \frac{AB}{BC} = \frac{c}{a}$$

$$\cot A = \frac{AC}{BC} = \frac{b}{a} \qquad \cot B = \frac{BC}{AC} = \frac{a}{b}$$

As the angles of a triangle add up to 180°, and angle C is a right-angle, it follows that $A + B = 90°$. A and B are said to be **complementary** angles.

The following are some of the more important relationships arising from the above:

$$\tan A = \frac{\sin A}{\cos A} \qquad \cot A = \frac{\cos A}{\sin A}$$

$$\sin A = \cos (90 - A) \qquad \cos A = \sin (90 - A)$$

$$\tan A = \cot (90 - A) \qquad \cot A = \tan (90 - A)$$

The theorem of Pythagoras

This theorem gives us the relationship that the square of the hypotenuse is equal to the sum of the squares of the other two sides. Thus in fig. 17, for the triangle ABC, $a^2 + b^2 = c^2$.

Considering areas in fig. 17:

$$\text{large square} = \text{small square} + 4 \text{ triangles}$$
$$(a + b)^2 = c^2 + 4(\tfrac{1}{2}ab)$$
$$a^2 + 2ab + b^2 = c^2 + 2ab$$
$$\therefore \quad \mathbf{a^2 + b^2 = c^2} \qquad (1)$$

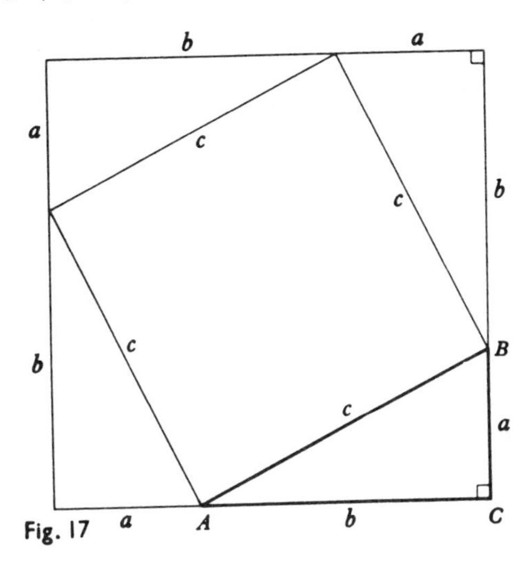

Fig. 17

If now we divide each side by c^2 we get,

$$\left(\frac{a}{c}\right)^2 + \left(\frac{b}{c}\right)^2 = 1$$

i.e. $$\mathbf{\sin^2 A + \cos^2 A = 1} \qquad (2)$$

If each side of the equation had been divided by b^2,

$$\left(\frac{a}{b}\right)^2 + 1 = \left(\frac{c}{b}\right)^2$$

i.e. $$\mathbf{\tan^2 A + 1 = \sec^2 A} \qquad (3)$$

If each side of the equation had been divided by a^2,

$$1 + \left(\frac{b}{a}\right)^2 = \left(\frac{c}{a}\right)^2$$

i.e. $$\mathbf{1 + \cot^2 A = \text{cosec}^2 A} \qquad (4)$$

Some important triangles

Set squares are usually triangles with angles either 45°, 45° and 90°, or 30°, 60° and 90°, i.e. either half a square or half of an equilateral triangle.

In fig. 18(a) we have a square with a side of unit length, and it is required that we should find the length of the diagonal AC.

Calling this length d units, we have by the theorem of Pythagoras:

$$d^2 = 1^2 + 1^2$$
$$d^2 = 2$$
$$d = \sqrt{2}$$

Therefore we can now write down our trigonometrical ratios for 45°:

$$\sin 45^\circ = \frac{1}{\sqrt{2}} \qquad \cos 45^\circ = \frac{1}{\sqrt{2}} \qquad \tan 45^\circ = 1$$

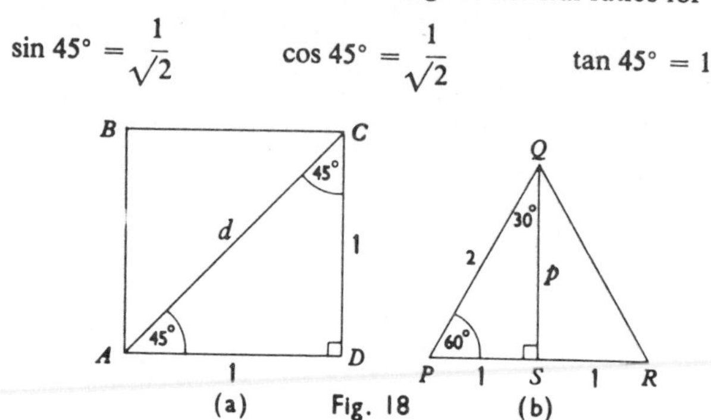

(a) Fig. 18 (b)

In fig. 18(b) we have an equilateral triangle of side 2 units symmetrically divided into two equal parts and it is required that we should find the length of the perpendicular QS:

Calling this length p units, we have by the theorem of Pythagoras:

$$2^2 = 1^2 + p^2$$
$$p^2 = 3$$
$$p = \sqrt{3}$$

Therefore we can now write down our trigonometrical ratios for 30° and 60°:

$$\sin 30^\circ = \frac{1}{2} \qquad \cos 30^\circ = \frac{\sqrt{3}}{2} \qquad \tan 30^\circ = \frac{1}{\sqrt{3}}$$

$$\sin 60^\circ = \frac{\sqrt{3}}{2} \qquad \cos 60^\circ = \frac{1}{2} \qquad \tan 60^\circ = \sqrt{3}$$

It is often convenient to form a right-angled triangle the sides of which have lengths that are whole numbers. Such proportions may be obtained from the relationship

$$(a^2 - b^2)^2 + (2ab)^2 = (a^2 + b^2)^2$$

by giving suitable whole number values to a and b.

The combination most frequently used in building for setting out right-angles accurately is the 3 : 4 : 5 relationship, which has been known and used for thousands of years. Other whole number combinations will be found in exercise 5(a)

APPLICATIONS OF TRIGONOMETRICAL RATIOS

1. Solution of right-angled triangles

(a) Given two sides, the remaining side may be found by applying the theorem of Pythagoras, and the angles by using sine, cosine or tangent as appropriate.

(b) Given one side and one angle, the remaining angle may be found from the fact that the angles are complementary, and the sides by using sine, cosine or tangent as appropriate.

Example 1 In the triangle ABC, right-angled at C, AB is 40 m and BC is 9 m. Find AC and the angles at A and B.

We first draw a small diagram and insert all the given information (fig. 19). From the theorem of Pythagoras:

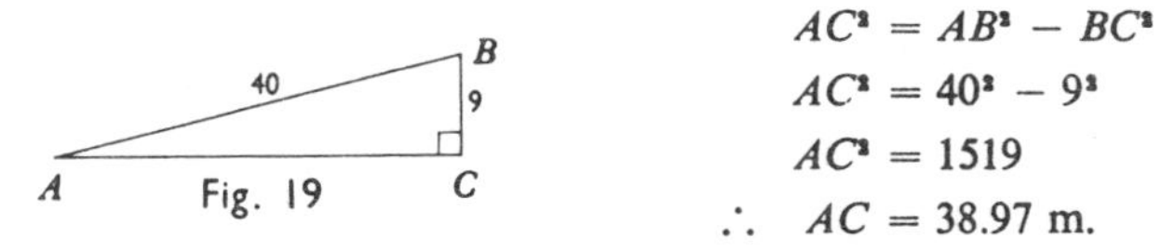

Fig. 19

$$AC^2 = AB^2 - BC^2$$
$$AC^2 = 40^2 - 9^2$$
$$AC^2 = 1519$$
$$\therefore \quad AC = 38.97 \text{ m.}$$

To find angle A, note first that the two given sides are in the positions of hypotenuse and opposite relative to $\angle A$, so we use the sine.

$$\sin A = \frac{\text{opposite}}{\text{hypotenuse}} = \frac{BC}{AB} = \frac{9}{40} = 0.225$$

Using a table of natural sines we see that

$$\sin 13^\circ = 0.2250$$
$$\therefore \quad \angle A = 13^\circ$$

To find angle B, note that the two given sides are in the positions of hypotenuse and adjacent relative to $\angle B$, so we use the cosine:

$$\cos B = \frac{\text{adjacent}}{\text{hypotenuse}} = \frac{BC}{AB} = \frac{9}{40} = 0.225$$

From a table of natural cosines we have:

$$\cos 77^\circ = 0{\cdot}225$$
$$\therefore \quad \angle B = 77^\circ$$

Check: A and B should be complementary:

$$13^\circ + 77^\circ = 90^\circ$$

Note that we could have used the calculated length of AC to find either of the angles, but then, had there been an error in calculating AC, it would have affected the result for the angles. It is always better to work as long as possible directly from the given information.

Example 2 Triangle PQR has an angle of $38\frac{1}{2}^\circ$ at P and a right-angle at R. Given that PQ is 16 m, find the lengths of the other two sides.

We first draw a small diagram and insert all the given information.

To find QR, note that QR and the given side PQ are in the positions of opposite and hypotenuse relative to the given angle P, so we use the sine.

$$\sin P = \frac{\text{opposite}}{\text{hypotenuse}} = \frac{QR}{PQ}$$

$$QR = PQ \times \sin P$$
$$= 16 \times 0.6225 \text{ m}$$
$$= 9.96 \text{ m.}$$

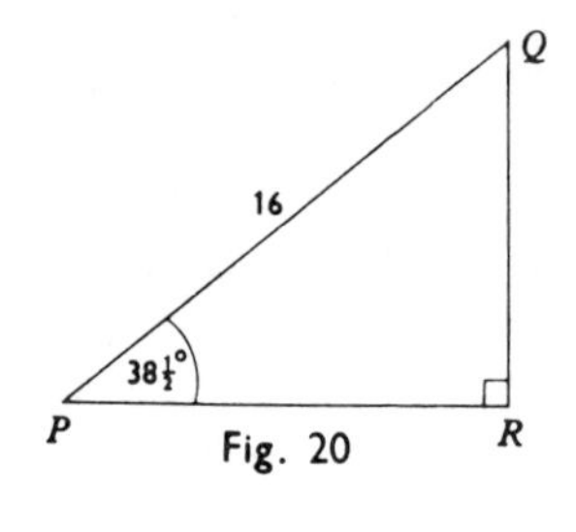

Fig. 20

Similarly, to find PR we use the cosine:

$$\cos P = \frac{\text{adjacent}}{\text{hypotenuse}} = \frac{PR}{PQ}$$

$$PR = PQ \times \cos P$$
$$= 16 \times 0.7826 \text{ m}$$
$$= 12.52 \text{ m.}$$

Example 3 ABC is a symmetrical roof truss, the rafters AB and BC each of length 4 m being inclined at an angle of 40° to the horizontal. Calculate the span and rise.

Fig. 21

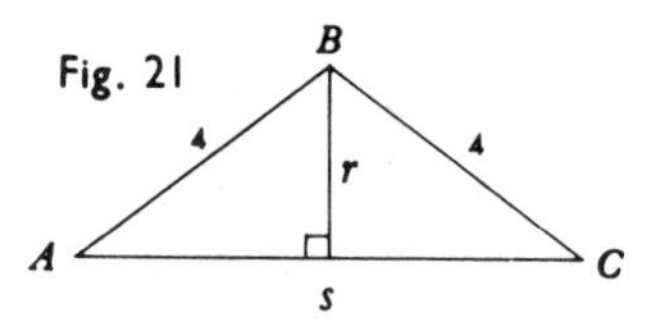

The roof truss is shown in fig. 21. As it is symmetrical it will be sufficient to calculate the dimensions of one-half of it. A perpendicular has been dropped from B on to AC. Denoting the rise by r, and the span by s, we have:

$$\sin 40° = \frac{r}{4} \quad \text{and} \quad \cos 40° = \frac{\frac{1}{2}s}{4}$$

$$r = 4 \sin 40° \qquad s = 8 \cos 40°$$
$$r = 4 \times 0.6428 \qquad s = 8 \times 0.7660$$
$$r = 2.57 \text{ m} \qquad s = 6.13 \text{ m.}$$

II. Gradients

If two points are selected on a straight-line graph, then the gradient (or slope) is given by the difference in the y-values (ordinates) divided by the difference in the x-values (abscissae).

$$m = \frac{y_1 - y_2}{x_1 - x_2}$$

and this was stated in chapter 2 to be equal to tan θ, where θ is the angle the line makes with the x-axis. This is easily seen, as the ratio is $\frac{\text{opposite}}{\text{adjacent}}$.

Example 4 Find the angle made with the x-axis by the line through the points (4, 5) and (− 1, − 3).

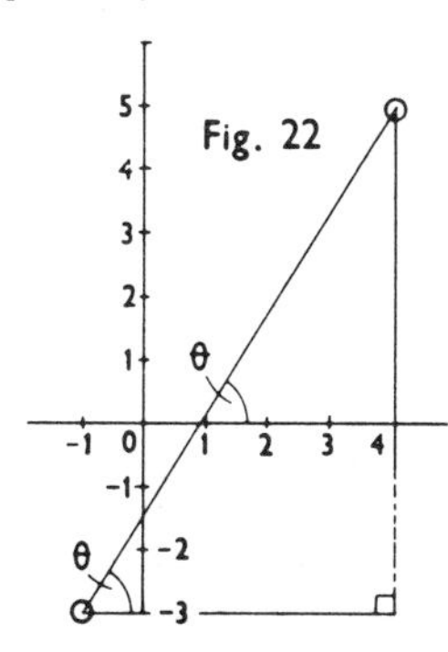

The line and the angle are shown in fig. 22.

$$\tan \theta = \frac{5 - (-3)}{4 - (-1)} = \frac{8}{5} = 1.6$$

$$\therefore \quad \theta = 58°$$

The pitch of a roof is given by the ratio $\frac{\text{rise}}{\text{span}}$ irrespective of whether it is a span roof or a lean-to type.

Example 5 A symmetrical span roof has a pitch of 1/5. Calculate the angle of slope.

For the tangent of the angle of slope, we need the rise over half the span. Thus the tangent of the angle of slope α is given by:

$$\tan \alpha = \frac{1}{2\frac{1}{2}} = 0.4000 \qquad \therefore \quad \alpha = 21° \, 48'$$

III. Deduction of trigonometrical ratios one from another

Method 1. Look up the given ratio in the tables to find the angle, then refer to the appropriate tables to find the new ratio.

Example: Given $\sin x = 0.55$, find $\tan x$.
From sine tables: $\sin 33° \, 22' = 0.5500$ $\quad \therefore \quad x = 33° \, 22'$
From the tangent tables: $\tan x = 0.6586$

Method 2. Draw a small diagram, insert the given ratio, by the theorem of Pythagoras, find the third side and hence the required ratio.

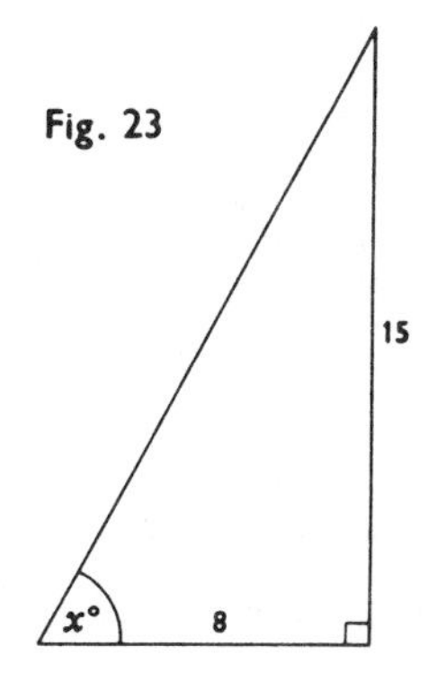

Example: Given $\tan x = 1\frac{7}{8}$, find $\cos x$.

$$\tan x = \frac{\text{opposite}}{\text{adjacent}} = \frac{15}{8}$$

These values are put in as shown in fig. 33.
Now $15^2 + 8^2 = 225 + 64 = 289 = 17^2$
$\therefore$ the hypotenuse is 17, and

$$\cos x = \frac{8}{17}$$

Method 3. Use the relationships $\sin^2 A + \cos^2 A = 1$ and $\tan A = \dfrac{\cos A}{\sin A}$.

Example: Given $\sin A = \frac{3}{5}$, find $\cos A$, and $\tan A$.

$$\cos^2 A = 1 - \sin^2 A$$

$$= 1 - \frac{9}{25} = \frac{16}{25}$$

$$\therefore \quad \cos A = \frac{4}{5}$$

$$\tan A = \sin A \div \cos A = \frac{3}{5} \div \frac{4}{5}$$

$$\therefore \quad \tan A = \frac{3}{4}$$

Angles of elevation and depression

The meaning of these commonly used terms is fully apparent from fig. 24, in which $x°$ = angle of elevation of B from A, $y°$ = angle of depression of A from B, the angles being measured from the horizontal.

Note that $x = y$.

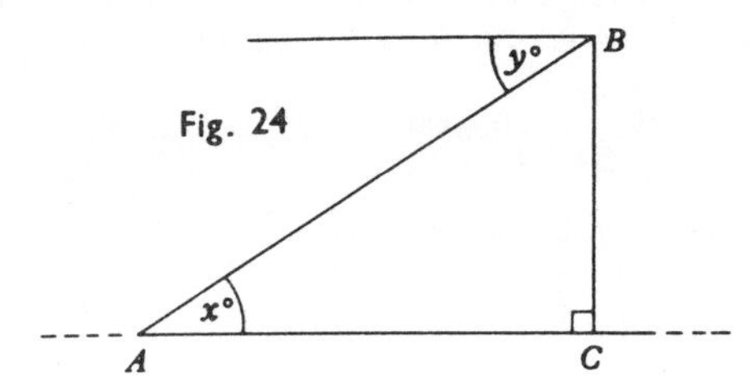

Fig. 24

EXERCISE 5(a)

1. ABC is a triangle right-angled at C. Given that AC is 12 m and BC is 5 m find AB and angles A and B.
2. Find the smallest angle and the hypotenuse of a right-angled triangle in which the other two sides are 35 m and 12 m.
3. Find the smallest side and smallest angle in a right-angled triangle given that the remaining sides are 40 m and 41 m.
4. If the foot of a ladder is placed on level ground $3\frac{1}{2}$ m from a vertical wall and it just touches the wall at a point 12 m above the ground, find the length to which the ladder is extended.
5. A vertical flagpole casts a shadow of 8 m on level ground when the angle of elevation of the sun is 49°36′. What is the height of the flagpole?
6. The diagonal of a rectangle is 1402 mm and each long side is 1302 mm. Find the length of a short side.

7. A man travels 400 m up a steady slope of 16°. How much higher is he now than when he started?

8. A lean-to garage roof has a span of $2\frac{1}{2}$ m and a rise of 1 m. Find the angle of slope of the roof.

9. Two straight lines on a graph cross one another at (1, 2). One line also passes through (−4, −6). The other line passes through (−9, −5). Find the angle each line makes with the axis and hence prove that the angle at which they cross one another is 23°.

10. If $\sin x = \frac{1}{3}$, find the value of $\cos x$ and $\tan x$ without using trigonometrical tables. (U.L.C.I.)

11. If $\tan A = \frac{39}{80}$, find the value of $\cot A$, $\cos A$ and $\operatorname{cosec} A$ without using trigonometrical tables.

12. A symmetrical span roof has rafters of 4 metres and a span of 7 metres. Find the rise and the angle of slope.

IV. Projection

Considering again a triangle ABC right-angled at C as in fig. 25, the adjacent side AC represents a plan view of the hypotenuse AB. It gives the projection of AB on to the horizontal plane. Similarly, BC is the projection of AB on to the vertical plane.

Now $\dfrac{AC}{AB} = \cos A$, and $\dfrac{BC}{AB} = \sin A$.

Therefore

$$AC = AB \cos A \quad \text{and} \quad BC = AB \sin A$$

or $$b = c \cos A \quad \text{and} \quad a = c \sin A$$

Fig. 25

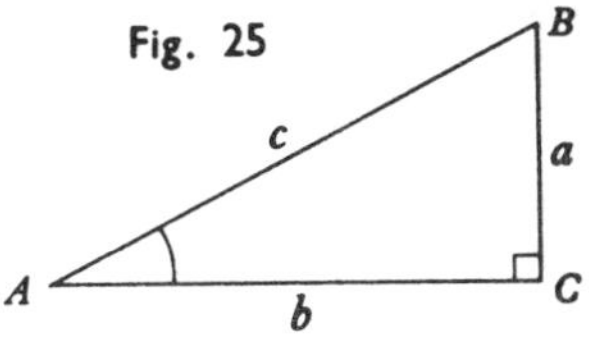

V. Angles of any magnitude

So far we have only considered acute angles, i.e. angles less than 90°; we must also be able to find the trigonometrical ratios of obtuse, reflex, and sometimes even negative angles.

Consider a circle of unit radius as shown in fig. 26. Since OP, the hypotenuse of triangle OAP, is of unit length, its projections on the horizontal and vertical axes through O are simply $\cos x$ and $\sin x$ respectively. Thus for any angle x, the length OA on the horizontal axis would give the cosine of the angle, and the length OB on the vertical axis would give the sine of the angle.

Consider the position of OP when the angle x is obtuse. Then, for an angle between 90° and 180°, the triangle will be in the second quadrant of the circle, so, although the projection for $\sin x$ will be on the positive vertical axis, the projection for cosine will be on the *negative* horizontal axis, thus showing that cosines of all obtuse angles are negative.

Further anticlockwise rotation of OP will give reflex angles, and it can be seen that in the third quadrant, 180° to 270°, both sine and cosine will be negative, whilst in the fourth quadrant, 270° to 360°, the sines are negative and the cosines positive.

Negative angles are given by rotating clockwise from the zero position, and angles greater than 360° are given by rotating a complete revolution plus the amount by which the angle exceeds 360°.

For tangents, consider fig. 27: $\tan x$ in triangle OQR is $\dfrac{QR}{OR}$, but $OR = 1$ for the circle of unit radius, therefore $\tan x$ is represented by the length of the intercept QR on the tangent to the circle at R. For angles in the first and third quadrants the tangents are positive, whereas for angles in the second and fourth quadrants the tangents are negative.

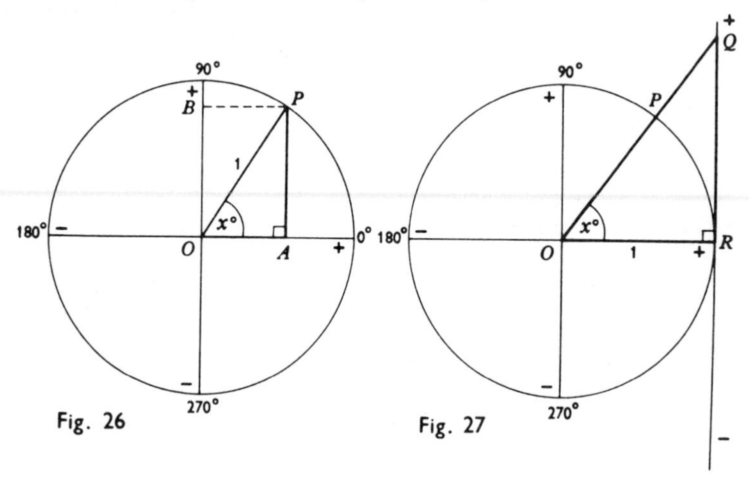

Fig. 26

Fig. 27

To find the value of the sine, cosine or tangent of any angle greater than 90°, first fix its sign as above, then refer to the angle symmetrically placed in the first quadrant and look up the value of this in the appropriate table.

The diagram of fig. 28 indicates which trigonometrical ratios are positive in each quadrant.

Examples:

(1) To find sine, cosine and tangent of 140°. From the composite diagram of fig. 29, we see that the sine is positive, whilst the cosine and tangent are negative. The corresponding angle in the first quadrant is $180° - 140° = 40°$,

$$\therefore \quad \sin 140° = \sin 40° = 0.6428$$

$$\cos 140° = -\cos 40° = -0.7660$$

$$\tan 140° = -\tan 40° = -0.8391$$

(2) To find the angles between 0° and 360° which have a sine of -0.6. Referring to a table of natural sines, we see that $\sin 36° 52' = +0.6$.

From fig. 30 we see that the two angles which have a sine of -0.6 are $(180° + 36° 52')$ and $(360° - 36° 52')$, i.e. $216° 52'$ and $323° 8'$.

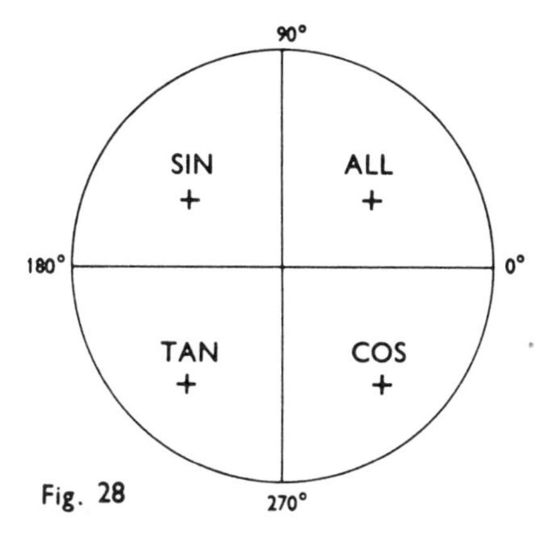

Fig. 28

Graphs of trigonometrical functions

Using the above system for finding the values of trigonometrical ratios for angles of any magnitude, we can now work out suitable sets of values and plot the graphs. The graphs of $y = \sin x$, $y = \cos x$ and $y = \tan x$ for values of x between 0° and 360° are shown in figures 31, 32 and 33.

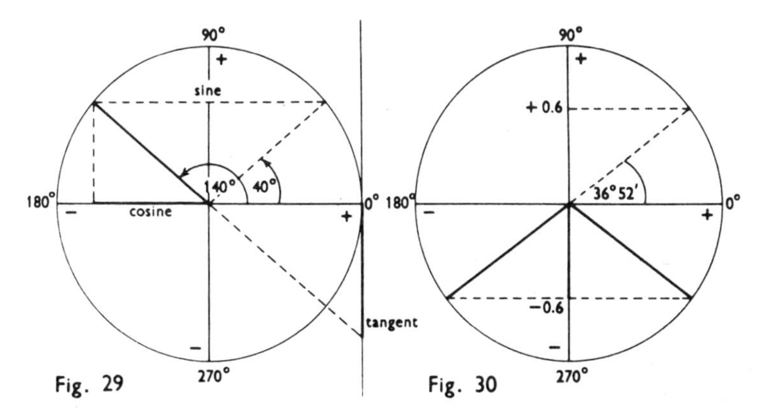

Fig. 29

Fig. 30

It can be seen that the graphs of $\sin x$ and $\cos x$ are continuous and they repeat at intervals of 360°. The graph of $\tan x$ is discontinuous and repeats at intervals of 180°.

These graphs form a convenient method for determining the angles corresponding to a given trigonometrical ratio. In the example above, angles with a sine of -0.6 were required: reference to the sine graph shows that a line drawn across at -0.6 would intersect the graph at two places, approximately 217° and 323°. Reference to a table of natural sines enables the results to be determined more accurately.

Inverse ratios

When we want to refer to an angle with a sine of 0.5, we may use the convenient notation arc sin 0.5. Similarly, arc tan 3 means the angle whose tangent is 3. Alternatively, these can also be written as $\sin^{-1} 0.5$ and $\tan^{-1} 3$, but this notation is not recommended for use with SI units (to avoid any possible confusion with reciprocals).

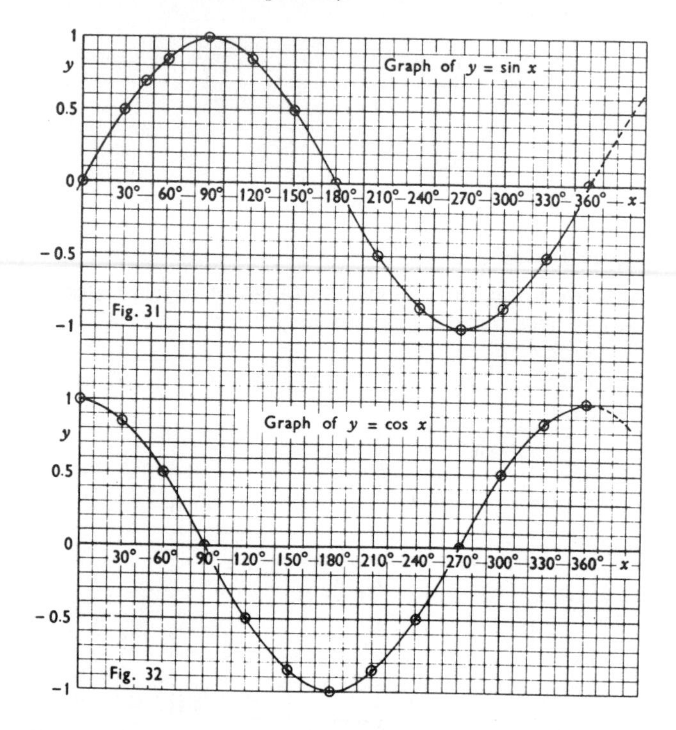

Fig. 31

Fig. 32

VI. Simple trigonometrical equations

These equations may be solved in exactly the same way as you would solve the usual type of linear or quadratic equation except that the unknown may be $\sin x$ or $\cos x$ or some other trigonometrical ratio, and the value (or values) of the angle x must finally be determined. Usually all possible solutions between 0° and 360° are required.

Examples:

(1) Find all values of x between 0° and 360° which satisfy the equation

$$\sin^2 x = 4 \cos^2 x$$

Divide each side by $\cos^2 x$:

$$\frac{\sin^2 x}{\cos^2 x} = \tan^2 x = 4$$

$$\tan x = \pm 2$$

$\tan x = +2$ gives solutions $63° 26'$ and $180° + 63° 26'$

$\tan x = -2$ gives solutions $180° - 63° 26'$ and $360° - 63° 26'$.

The complete solution is therefore

$$x = 63° 26',\ 116° 34',\ 243° 26' \text{ or } 296° 34'$$

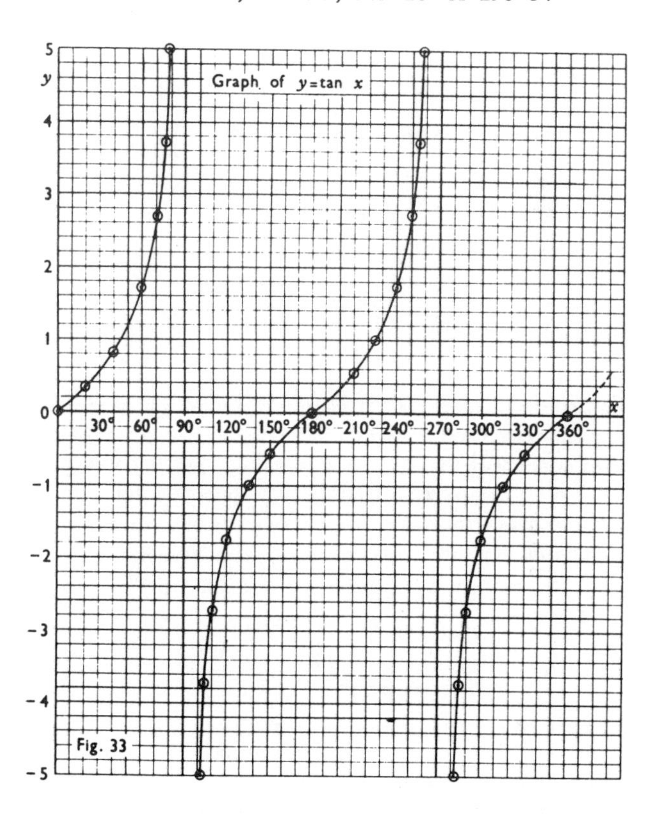

Fig. 33

(2) Solve the following equation for values of θ between 0° and 360°.

$$\tan\theta + 3\sin\theta = 0$$

Now $\tan\theta = \dfrac{\sin\theta}{\cos\theta}$

Factorising, $\sin\theta\left(\dfrac{1}{\cos\theta} + 3\right) = 0$

Therefore, either

$$\sin\theta = 0 \text{ or } \frac{1}{\cos\theta} + 3 = 0 \text{ i.e. } \cos\theta = -\frac{1}{3}$$

either

$$\theta = 0^\circ,\ 180^\circ,\ 360^\circ \text{ or } 109^\circ\,28' \text{ or } 250^\circ\,32'$$

(3) Solve, for values of x between 0° and 360° the equation:

$$6\sin^2 x + \cos x = 4$$

Now this equation as it stands contains two variables, $\sin x$ and $\cos x$, we therefore use the relation $\sin^2 x = 1 - \cos^2 x$ to bring it to a quadratic in $\cos x$.

$$6(1 - \cos^2 x) + \cos x = 4$$

$$6\cos^2 x - \cos x - 2 = 0$$

Factorising,

$$(3\cos x - 2)(2\cos x + 1) = 0$$

$$\cos x = \tfrac{2}{3} \text{ or } \cos x = -\tfrac{1}{2}$$

$$x = \left.\begin{matrix}48^\circ\,11' \\ 311^\circ\,49'\end{matrix}\right\} \text{ or } \left.\begin{matrix}120^\circ \\ 240^\circ\end{matrix}\right\}$$

Circular measure

Angles can either be measured in degrees, minutes and seconds, or alternatively, in radians and milliradians. The relation between the units is that π radians are equivalent to 180°, i.e. 1 radian $= 180^\circ/\pi$, or approximately $57^\circ\,18'$. A complete revolution of 360° is 2π radians and this is an advantage in the calculation of lengths of arc or areas of parts of a circle (see chapter 6).

For small angles, measurements are made in milliradians (1000 mrad = 1 rad). Conversions can be made by using the following relationships:

$$1^\circ = 17.45 \text{ mrad} \qquad 1 \text{ mrad} = 3'\,26''.$$

EXERCISE 5(b)

1. Find the sines, cosines and tangents of the following angles:
   (a) $73^\circ\,49'$ (b) $100^\circ\,10'$
   (c) $200^\circ\,20'$ (d) $333^\circ\,33'$

2. Find the sines, cosines and tangents of:
   (a) 500° (b) -52°

3. Evaluate:
   (a) arc sin $\tfrac{1}{3}$ (b) arc cos $\tfrac{2}{3}$
   (c) arc tan 2.500

4. Draw the graph of $y = \sin x + \cos x$ and hence find the maximum and minimum values of $\sin x + \cos x$ over the range 0° to 360° for x.

5. Find all the values of x between 0° and 360° which satisfy the equations:
   (a) $9\sin^2x = 4$ (b) $\cos^2x + 2\sin^2x = 1.7$
   (c) $3\sin^2x = 15\cos^2x$

6. Solve the following equations for values of θ between 0° and 360°
   (a) $5\sin\theta - 2\tan\theta = 0$ (b) $4\cos\theta = 3\cot\theta$

7. Solve, for values of x between 0° and 360°, the equations:
   (a) $9 + 6\sin x - 8\cos^2x = 0$ (b) $1 + \cos x = \frac{4}{5}\sin^2x$
   (c) $\sec^2x - 3\tan x - 5 = 0$

8. Express the following angles in degrees:
   (a) $\frac{\pi}{4}$ rad (b) $\frac{7\pi}{12}$ rad
   (c) 1.2π rad (d) 5π rad

REVISION EXERCISE 5

1 (a) Show that $\cos^2\theta + \sin^2\theta = 1$
   (b) Solve the following equations for values of θ between 0° and 360°.
      (i) $2\tan^2\theta - \tan\theta - 1 = 0$
      (ii) $2\cos^2\theta - 1 = 0$ (K.C.E.A.B.)

2. Find the cosine and cosecant of the following angles:
   (a) 4 rad
   (b) $1.8\,\pi$ rad
   (c) 150 mrad

3. (a) Prove $\sin^2A + \cos^2A = 1$, where A is an acute angle in a right-angled triangle ABC.
   (b) Simplify:
   $$\frac{\cos^2x(1 - \sec^2y)\sin x}{(1 - \sin^2x)\cos x\tan^2y}$$ (U.L.C.I.)

4. In a triangle ABC, angle CAB is 35° and angle CBA is 55°. If $AB = 3.2$ m, find AC, BC, and the length of the perpendicular from C onto AB.

5. Plot the graph of $3\sin x + 4\cos x$ for values of x from 0° to 180° and hence find the maximum value of y and the corresponding value of x. (E.M.E.U.)

6. If $a\cos\theta = 2$ and $a\sin\theta = 3$, calculate a and θ and check your results by means of a diagram. (E.M.E.U.)

7. (a) Using appropriate tables, give the values of:
      (i) sin 215° (ii) cos 320° (iii) tan 160°
   (b) A steel chimney is held in position by several guy wires. Two such wires are AB and AC where B, C and the base of the chimney, D, are on the same level in a straight line. If angle $ABC = 42°\,10$, $ACD = 67°\,25'$, $BC = 10$ m, find the height of A above D. (N.C.T.E.C.)

8. Set up a table of values of y for values of θ between 0° and 360°, at 30° intervals, for the equation:

$$y = 1 + 2 \cos \theta + \cos 2\theta$$

Draw the graph of the relationshlp and give the values which satisfy the condition $y = -0.25$. (N.C.T.E.C.)

9. A man is walking up a slope and notes that for 105 m up the slope he is going at an inclination of 15° with the horizontal, and for a further 55 m the inclination is 12°. Find the total distance travelled horizontally and the total height attained. (U.L.C.I.)

10. A straight railway line runs at a steady slope between two villages 5 km apart. If the angle at which the track slopes is 3.5°, find the difference in level between the two villages.

11. A mine tunnel slopes downwards at an angle of 25° to the horizontal for a distance of 1·6 km measured on plan. Find (a) how far the end of the tunnel is below the surface of the earth; (b) the distance a man must travel down the tunnel, measured on the slope, to be 800 m below the surface. (U.L.C.I.)

12. A tower stands on level ground, and from a point some distance away the angle of elevation of the nearest corner is 27°. Because of an intervening moat, the distance to the base of the tower cannot be measured directly, but from a point 20 m nearer to the tower, the angle of elevation is 34°. Estimate the height of the tower.

13. (a) Convert

    (i) 225° to radians, giving your answer in terms of π,

    (ii) $\frac{4\pi}{9}$ radians to degrees.

    (b) Find two values of A between 0° and 360° such that $\sin A = -0.766$.

    (c) If $\sin A = \frac{3}{5}$ find without tables $\cos A$ and $\tan A$. (E.M.E.U.)

14. (a) The tangent of an acute angle is $\frac{12}{5}$. Without using tables find the value of the sine and cosine of the angle.

    (b) Without using tables evaluate $2 \sin 60° \tan 30°$. (E.M.E.U.)

15. (a) By use of tables or otherwise find the value of one radian in degrees and minutes correct to the nearest minute.

    (b) Express 40° in radians correct to three decimal places. (E.M.E.U.)

16. Plot the graph of $y = 3 \sin x + 2 \sin 2x$ between $x = 0°$ and $x = 360°$ and state the values of x which give maximum and minimum values of y. What are these maximum and minimum values? (N.C.T.E.C.)

17. (a) Prove the identity:

$$1 + \tan^2 A = \sec^2 A$$

    (b) Evaluate:

$$\frac{\sin 133° 13' - \tan 281° 42'}{\cos 215° 17' - \sec 330° 14'}$$

(U.L.C.I.)

18. (a) Find without tables:

    (i) $\tan 135°$ (ii) $\cos 240°$ (iii) $\sin 330°$

    (b) Prove that:

$$\sin^4\theta - \cos^4\theta = 1 - 2 \cos^2\theta$$

(U.L.C.I.)

19. Plot the two graphs $y = \sin\theta$ and $y = \cos\theta$ using the same axes. Let θ vary in each case from 0° to 90°. Find from the graphs a solution to these simultaneous equations. (U.L.C.I.)

20. (a) Find the value of: $\cos 210° \cos 240° + \sin 330° \cos 30°$.

(b) If $\sin\theta = \dfrac{x^2 - y^2}{x^2 + y^2}$, find the values of $\cos\theta$ and $\cot\theta$. (U.L.C.I.)

21. (a) If $\sin A = \dfrac{5}{13}$ and $\cos B = \dfrac{7}{25}$ calculate, without the use of tables:

$$\cos(90 - A)\sin(90 - B)\tan(180 + A)\cot(270 + B)$$

(b) Prove:

$$(\sec y - \cos y)(\operatorname{cosec} y - \sin y) = \sin y \cos y$$ (U.L.C.I.)

22. (a) Without using tables, find the value of $4\sin\theta + \cos\theta$, when:

$$\theta = 30°,\ 120°,\ 240°,\ \text{and } 300°$$

(b) Prove:

$$\sin\theta\cos\theta\cot\theta\sec\theta = \cos\theta$$ (U.L.C.I.)

23. Prove that:

$$[\sin^2 A + \cos^2 A + \tan^2 B]\sin^2(90 - B) = 1$$ (U.L.C.I.)

24. Find the values of θ, less than 180°, which satisfy the following equations:

(a) $5\tan^2\theta - \sec^2\theta = 11$;

(b) $\tan\theta + \cot\theta = 2$. (U.L.C.I.)

25. Find the values of x and y between 0° and 360° which satisfy the following simultaneous equations:

$$3\sin x - 2\cos y = 1$$
$$\sin x + \cos y = 1$$ (U.E.I.)

6 Areas and Volumes

CALCULATIONS involving distances, areas and volumes have been simplified by the introduction of SI units and the consequent abolition of duodecimals, etc. No doubt future generations will wonder why we continued so long with a system involving such awkward units as miles, yards and inches!

The use of SI units

For linear measure we have now accepted the use of the **metre** (m), whilst for long distances we use the **kilometre** (km) which is 1000 times larger, and for small dimensions the **millimetre** (mm) which is 1000 times smaller. Each of these gives rise to corresponding units for area and volume, but it is essential to understand that the 1000:1 ratio is no longer maintained, e.g. 1 km^3 is actually a short form of $1(km)^3$ and it follows therefore that

$$1\ km^3 = 1000\ m \times 1000\ m \times 1000\ m$$

i.e.

$$1\ km^3 = 10^9\ m^3$$

The units and their relationships can be summarised as follows:

LENGTH	AREA	VOLUME
kilometre	square kilometre	cubic kilometre
metre	square metre	cubic metre
millimetre	square millimetre	cubic millimetre
$1\ km = 10^3\ m$	$1\ km^2 = 10^6\ m^2$	$1\ km^3 = 10^9\ m^3$
$1\ m = 10^3\ mm$	$1\ m^2 = 10^6\ mm^2$	$1\ m^3 = 10^9\ mm^3$

Because the cubic millimetre is so very small, and the cubic metre rather large, intermediate units have been introduced for liquid measure. Thus we now have 1000 litres to one cubic metre, 1000 millilitres = 1 litre, and 1000 mm^3 = 1 ml.

Example: Calculate the quantity of water in a flat open trough 4 m long, and 750 mm wide when the depth of water is 150 mm.

$$\begin{aligned} \text{Volume} &= 4\ m \times 750\ mm \times 150\ mm \\ &= 4\ m \times 0.75\ m \times 0.15\ m \\ &= 0.45\ m^3 \\ &= 450\ \text{litres} \end{aligned}$$

Mass and Weight

The weight of a given mass is dependant upon gravity. An astronaut who is walking on the surface of the moon is conscious of the fact that his weight is only one sixth of his weight upon Earth, although his mass is unchanged. When we are concerned with calculating such things as the mass of concrete in a solid block, we need only find the product of volume and density. If the concrete block has then to be hoisted by a crane, and we need to know the force exerted by the block on the lifting chain, then we need the product of the mass and the value of the gravitational acceleration (g) at that particular site. The standard value of g is 9.806 65 m/s^2, but it is common practice to use 9.81 m/s^3 in construction problems. The difficulty is that the value of g varies slightly across the Earth's surface.

In the above illustration, the mass of concrete suspended from the crane is exerting a force on the chain which is holding it. This force is measured in **newtons** in the new system of units. Smaller forces are measured in **millinewtons** (mN) and larger forces in **kilonewtons** (kN). The basic unit of force, the newton, is defined as the force which, when applied to a body having a mass of one kilogram, gives it an acceleration of one metre per second squared.

Example: It is proposed that a section of wall should be built, provided that the existing hard surface can withstand the pressure. The wall is to be of brickwork, 8 m long, 1.5 m high, 215 mm thick. If the density of brickwork is 2250 kg/m^3, find the volume and mass of the wall and the pressure it will exert on the foundation.

$$\begin{aligned} \text{Volume} &= 8\text{ m} \times 1.5\text{ m} \times 215\text{ mm} \\ &= 8\text{ m} \times 1.5\text{ m} \times 0.215\text{ m} \\ &= 2.58\text{ m}^3 \\ \text{Mass} &= 2.58 \times 2250\text{ kg} \\ &= 5805\text{ kg} \\ \text{Thrust due to wall} &= 5805 \times 9.81\text{ N} \\ &= 56\,950\text{ N} \end{aligned}$$

Now,

$$\begin{aligned} \text{Pressure} &= \frac{\text{thrust}}{\text{area}} = \frac{56\,950\text{ N}}{8 \times 0.215\text{ m}^2} \\ &= 33\,110\text{ N/m}^2 \end{aligned}$$

Hence, in this case, the foundation must be capable of withstanding a pressure of at least 33 110 N/m^2 (33 kN/m^2)

Units of mass

In the example above, the mass was calculated in **kilograms** and this is the commonest unit for mass in the construction industry. It replaces the old units of lb and cwt quite easily. (kg = 2.2 lb, and 1 cwt is about 50 kg)

When dealing with smaller masses, as in laboratory tests, the smaller units of **grams** and **milligrams** are used. For very large masses, it is obvious that a larger unit than the kilogram is required. To fit in with the system it should be equivalent to 1000 kg. This larger unit of mass is called the **tonne.**

$$1\text{ g} = 10^3\text{ mg}$$
$$1\text{ kg} = 10^3\text{ g} = 10^6\text{ mg}$$
$$1\text{ t} = 10^3\text{ kg} = 10^6\text{ g} = 10^9\text{ mg}$$

EXERCISE 6(a)

1. Find the cost of two hardboard door panels, 2 m × 750 mm, at 27p/m².

2. A piece of timber for a floor joist is 50 mm × 175 mm × 3.2 m and its mass is 12.6 kg. Find the density of the timber.

3. A trench is to be excavated 5.6 m × 500 mm × 400 mm. If the average density of the earth to be removed is 1500 kg/m³, calculate the mass of earth involved.

4. Taking the density of masonry as 2500 kg/m³, calculate the pressure exerted on its foundation by a continuous, level stone wall 300 mm thick and 1600 mm high.

5. A rectangular stone block measures 1200 mm × 500 mm × 400 mm. If its density is 2600 kg/m³, find the force it exerts on the chain of the lifting tackle used to move it into position.

6. Show that the pressure of water with a head of 500 mm is 4905 N/m².

Triangles

The area of a triangle is given by ½ base × height, but it is not always possible to measure the height directly. One alternative is to measure 2 sides of the triangle and the angle between them (see fig. 34).

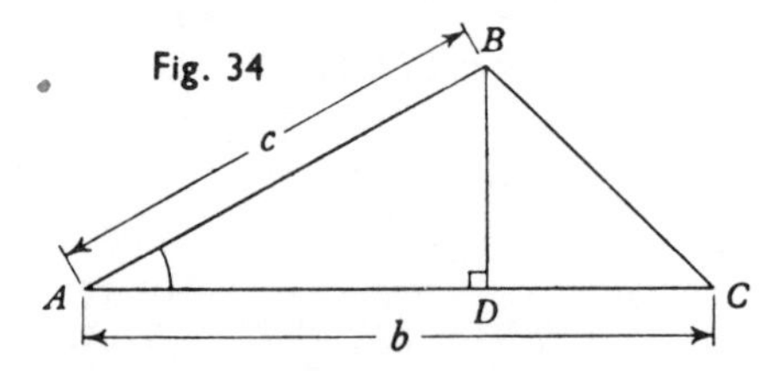

Considering BD as the vertical projection of AB, we have the height of the triangle, BD, given by $AB \sin A$, i.e. $c \sin A$. Multiplying this by the base length b and halving it for the area of the triangle, we have:

$$\text{area} = \tfrac{1}{2}bc \sin A$$

and similarly the alternative forms

$$\text{area} = \tfrac{1}{2}ab \sin C$$
$$\text{area} = \tfrac{1}{2}ac \sin B$$

Circular measure

In SI units, angles should be measured in radians. For various practical reasons, the use of the alternative system of degrees, minutes and seconds is still in use and a knowledge of both systems is therefore necessary.

As previously mentioned in chapter 5, 2π radians are equivalent to 360°. The circumference of a circle is 2π times the radius, and therefore a radian may be defined as the angle subtended at the centre of a circle by an arc equal in length to the radius. It follows from this definition of a radian that the length of arc which subtends an angle of θ radians at the centre of a circle is $r\theta$, where r is the radius. Thus in fig. 35 arc $PQ = r\theta$ and arc $RS = r\theta$.

Area of sector and segment

The area of a complete circle is πr^2 and this corresponds to a complete revolution of 360° or 2π radians. By proportion, a sector with an angle of θ radians must have an area of

$$\frac{\theta}{2\pi} \times \pi r^2 = \tfrac{1}{2}r^2\theta$$

To find the area of a segment (see fig. 35) we find the area of the sector and subtract from it the unwanted triangle. Using the formula for the area of the triangle, $\tfrac{1}{2}ab \sin C$, we have in this case

$$\text{area of triangle } ORS = \tfrac{1}{2}r^2 \sin\theta$$

$$\begin{aligned}\text{and thus the area of the segment} &= \text{sector} - \text{triangle}\\ &= \tfrac{1}{2}r^2\theta - \tfrac{1}{2}r^2\sin\theta\\ &= \tfrac{1}{2}r^2(\theta - \sin\theta)\end{aligned}$$

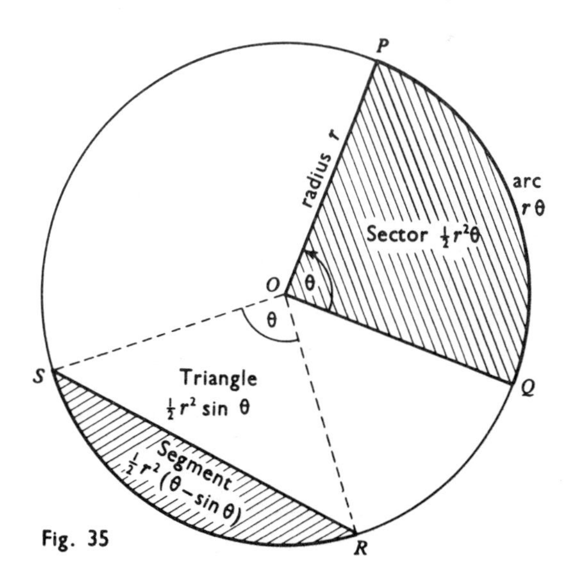

Fig. 35

Radius and angle of a segment

For the use of the above formula both the radius and the angle of the segment are required and must be derived from such measurements as can be made or may be supplied.

Example: Find the area of brickwork necessary to fill the space between the soffit and the springing line of a segmental arch of span 18 m and rise 3 m.

The radius may be found by applying the theorem of Pythagoras to the triangle *OBD* in fig. 36.

$$r^2 = (r - 3)^2 + 9^2$$

$$r^2 = r^2 + 6r + 9 + 81$$

$$6r = 90$$

$$r = 15 \text{ m}$$

Alternatively we may apply the geometrical theorem for intersecting chords:

$$AD.DB = CD.DE$$

$$9 \times 9 = (2r - 3)3$$

$$81 = 6r - 9$$

$$6r = 90$$

$$r = 15 \text{ m}$$

The angle of the segment is found by trigonometry from the same triangle *OBD*.

$$\sin \tfrac{1}{2}\theta = \tfrac{9}{15} = \tfrac{3}{5} = 0.6000$$

$$\tfrac{1}{2}\theta = 36° \ 52'$$

$$\theta = 73° \ 44' = 1.2869$$

$$\sin \theta = 0.9600$$

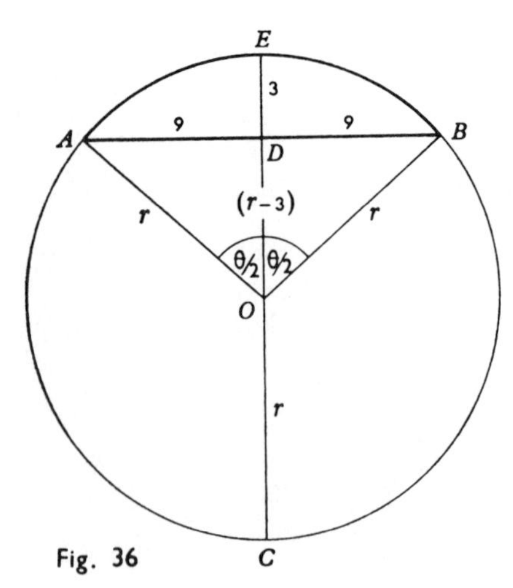

Fig. 36

now area

$$A = \tfrac{1}{2}r^2(\theta - \sin\theta)$$
$$= \tfrac{1}{2}(15)^2(1.2869 - 0.9600)$$
$$= 36.78 \text{ m}^2$$

Approximate formulae

To avoid having to calculate the angle at the centre, use is frequently made of certain approximate formulae. These only involve the measurements most easily made, i.e. the span (s), the rise (h), and the chord of half the arc (c). Referring to fig. 36, these would be as follows:

$$s = AB, \quad h = DE, \quad c = AE = EB$$

In terms of these measurements, Huyghens approximation to the length of arc (l) would be:

$$l = \frac{8c - s}{3}$$

For a segment the area (A) is given by:

$$A = \frac{h(4s^2 + 3h^2)}{6s}$$

or by

$$A = \frac{4h^2}{3}\sqrt{\frac{2r}{h} - 0.608}$$

For the example of the segmental arch previously considered, either of these two formulae would give the area with reasonable accuracy.

Thus, substituting in

$$A = \frac{h(4s^2 + 3h^2)}{6s}$$

$$A = \frac{3(1296 + 27)}{6 \times 18}$$
$$= 36.75 \text{ m}^2$$

Alternatively,

$$A = \frac{4h^2}{3}\sqrt{\frac{2r}{h} - 0.608}$$
$$= 12\sqrt{10 - 0.608}$$
$$= 12 \times 3.064$$
$$= 36.77 \text{ m}^2$$

We have thus three formulae for the area of a segment and the one to be used for a particular problem should be selected to correspond to the given measurements.

Any chord divides a circle into two segments, the larger known as the major segment, the smaller referred to as the minor segment. The formulae above are for a minor segment, i.e. for values of θ less than 180°.

EXERCISE 6(b)

1. Find the area of a rectangle 35 m × 12 m and also the length of a diagonal.

2. A house has a gable end 8 m wide, 10 m high to the ridge and 7 m high to the eaves. Find the surface area of this end face.

3. Find the number of rolls of wallpaper of length 10 m and width 500 mm which would be required to paper a ceiling 6500 mm × 5200 mm, allowing 10% for cutting, etc.

4. Find the area of a triangle ABC given that a = 320 mm, b = 250 mm and angle C is $37\frac{1}{2}$°.

5. $ABCD$ is a quadrilateral in which AB = 95 mm, BC = 150 mm and CD = 80 mm. If angle B is 90° and angle C is 112°, find the area of $ABCD$.

6. The area of triangle XYZ is 0.264 m². Given that XY = 820 mm and YZ = 690 mm, find angle Y.

7. In a certain parallelogram one side is 50 mm, another side is 45 mm, and one angle is 82°. Find the area of the parallelogram and also the perpendicular distances between each pair of parallel sides.

8. Find the area of a field in the shape of a parallelogram with adjacent sides of 65 m and 104 m and an included angle of 78°.

9. The depth of water in a culvert is uniform at 200 mm and it flows at 3 m/s. The width at the base is 1000 mm but 1200 mm at the water surface. Find the volume of water being carried in litres per second.

10. Find the length of arc which subtends an angle of 72° at the centre of a circle radius $10\frac{1}{2}$ m.

11. A circle has a diameter of 240 mm. Find the angle at the centre subtended by an arc length of 150 m.

12. Show that if l is the length of the arc of a sector of a circle radius r, then the area of the sector is $\frac{1}{2}rl$. Hence find the area of a sector radius 5 m. and arc length 9 m. Find also the angle of the sector.

13. Find the area of a segment of a circle radius 5.6 m for which the angle at the centre is 80°.

14. A chord of a circle radius 7 m passes exactly 1 m from the centre. Find the area of the two segments into which the circle is divided by the chord.

15. The cross-section of a passageway has its two vertical sides each 3 m high, the horizontal base 2 m long. The top is an arc of a circle. The centre of this circle is the centre of the rectangle formed by the base and sides. Find the area of the cross-section.

16. A segmental arch has a span of 16 m and a rise of 2 m. Find:
    (a) the radius of the arch;
    (b) the rise of the arch at a point on the springing line 3 m from one end.

17. In a circle of radius 100 mm a chord *AB* subtends an angle of 35° at the centre *O*. Calculate:
    (a) the area of the triangle *OAB*;
    (b) the area of the sector *OAB*; and
    (c) the area of the major segment formed by the chord *AB*. (U.L.C.I.)

REVISION EXERCISE 6

1. A rectangular storage tank has a base 1200 mm by 500 mm and contains liquid to a depth of 200 mm. Find the total value of the liquid in store at 8p per litre.

2. An ornamental stone sphere has a diameter of half a metre. If the density of the stone is 2640 kg/m^3, find the mass of the sphere.

3. Topsoil to a depth of 200 mm is to be stripped from a rectangular building site 20 m × 30 m. It is proposed that this soil should be stored in a conical pile with a base diameter of 10 m. Assuming that the soil bulks by 10% on excavation, show that the cone would rise to a height of just over 5 m.

4. In triangle *ABC*, *AB* = 120 mm, *AC* = 135 mm, and angle *A* is 50°.

   In triangle *XYZ*, *XY* = 220 mm, *XZ* = 60 mm, and angle *X* is 60°.

   Which triangle is the largest and what is the difference in area between them?

5. If the scale of a certain map is 1:1000, find the area in square metres of a plot of land shown on the map as a triangle with sides 24 mm and 25 mm separated by an angle of $65\frac{1}{2}$°.

6. Calculate the volume of an aircraft hangar, semi-circular in section, 50 m long, the arc length of the roof being 31.42 m. (U.L.C.I.)

7. A square duct, 200 mm × 200 mm, is delivering a current of air with a velocity of 4 m/s, into a hall whose dimensions are 20 m × 10 m × 8 m. Find, to the nearest minute, how long it will take for the air of the hall to be renewed. (U.L.C.I.)

8. The span of a segmental arch is $1\frac{1}{2}$ m and its rise $\frac{1}{2}$ m. Calculate:

   (a) the radius of the arch;

   (b) the area of the segment. (W.J.E.C.)

9. A cylindrical boiler 2.5 m long and 1.25 m diameter lies with its axis horizontal. Calculate the volume of water, in litres, required to fill the boiler to a depth of 1 m. (U.L.C.I.)

10. The cross-section of a tunnel is a major segment of a circle of diameter 8 m. The flat base subtends an angle of 60° at the centre of the circle. Calculate the volume to be excavated in the construction of a straight portion of the tunnel 100 m long. Take $\pi = 3.142$. (W.J.E.C.)

11. A copper hot-water tank is in the form of a cylinder with a hemispherical cap and stands on a flat base. The internal diameter of the cylinder is $\frac{1}{2}$ m and the overall internal height is 1 m. Find the capacity of the tank in litres.

12. Iron pipe of 100 mm internal diameter has a mass of 10 kg per metre. Taking the density of iron as 7.2 g/cm^3 (7.2 mg/mm^3), find the thickness of the pipe wall.

7 Irregular Areas and Volumes

NATURAL boundaries such as rivers and coastlines rarely follow straight lines or regular curves and therefore various methods are adopted for finding the areas of irregular figures.

By squared paper

The figure is drawn to scale on squared paper and the number of squares within the boundary of the figure estimated as accurately as possible. A large number of squares is required for accuracy and the method then becomes laborious.

By planimeter

A planimeter is an instrument for measuring plane areas. There are various types, but basically each has a pointer which is moved around the boundary of the area to provide automatically a measure of the area usually recorded on a graduated wheel or cylinder.

The Trapezoidal Rule

The area of any trapezoid is given by half the sum of the parallel sides times the perpendicular distance between them. Thus, if a base line be drawn across the area (preferably across its widest part) and the base line is then sub-divided into a number of equal parts and parallel lines drawn at right-angles to the base line through the divisions, the area is subdivided into a number of sections each of which may be regarded as being approximately trapezoidal in shape. If the parallel lines are of length $y_1, y_2, \ldots y_n$, the total area is given by

$$S[\tfrac{1}{2}(y_1 + y_n) + y_2 + y_3 + \ldots]$$

where S is the width of each strip.

It is convenient to remember this in the form:

Area = strip width × (average of first and last + sum of all the others)

This applies equally to ordinates on a graph, survey offsets, etc. Greater accuracy is obtained by increasing the number of strips, but if the curvature of the boundary is continually one way the estimate of the area can only be approximate.

The Mid-ordinate Rule

After the area has been divided into strips by equally spaced parallel lines as for the Trapezoidal Rule, the centre of each strip is measured. This represents an average height for each strip, and the area of a strip will be given by centre length (mid-ordinate) multiplied by the strip width. The total area will be given by the sum of all the mid-ordinates × strip width. It is convenient to mark off the lengths of the mid-ordinates consecutively along a straight edge of paper; in this way the measurements can be taken off the scale drawing very quickly and the total length multiplied by the strip width to give the area.

Simpson's Rule

This is the most accurate of the formulae: for many regular curves it is exact. The value obtained for a given area by the Trapezoidal Rule may err on one side, and then the value by the Mid-ordinate Rule would err on the other side, but the value by Simpson's Rule would lie in between the other two. In fact, the formula for Simpson's Rule may be obtained by combining one-third of the trapezoidal formula with two-thirds of the mid-ordinate formula.

The formula may be expressed as

$$\text{Area} = \frac{S}{3}(A + 4B + 2C)$$

where S = strip width,
A = sum of first and last ordinates,
B = sum of all even ordinates,
C = sum of remaining odd ordinates.

Note: This form of Simpson's Rule requires an *odd number of ordinates*, i.e. an *even number of strips*.

The formula may be conveniently remembered as

$$\text{Area} = \tfrac{1}{3}\text{ strip width} \times (\text{first} + \text{last} + TOFE)$$

where **TOFE** stands for Twice Odds + Four Evens.

Replacing ordinates by areas of cross-section enables volumes to be found by Simpson's Rule.

The Prismoidal Rule

This may be regarded as a modified form of Simpson's Rule with only three ordinates. The formula may be stated as

$$\text{Volume} = \frac{h}{6}(\text{base} + \text{top} + 4\text{ mid-section})$$

where h is the perpendicular distance between base and top; and the base, top, and mid-section are all parallel planes.

Example 1 Find the area enclosed by the curve $y = 9x + 36 - x^2$, and the positive x and y axes.

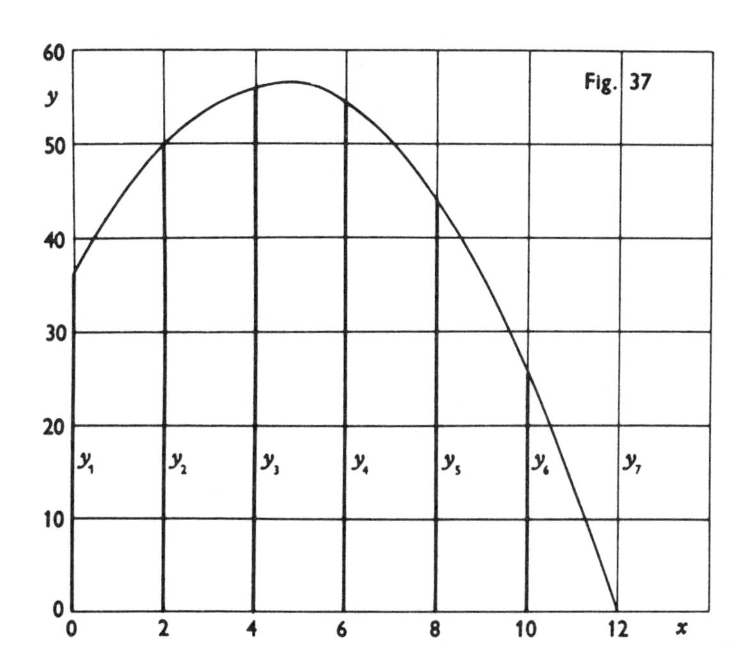

(a) By the Trapezoidal Rule:
Drawing up a table of values for even values of x—

x	0	2	4	6	8	10	12
y	36	50	56	54	44	26	0
	y_1	y_2	y_3	y_4	y_5	y_6	y_7

(A sketch of the graph is shown in fig. 37.)
By the Trapezoidal Rule:

$$\begin{aligned}\text{Area} &= S\left[\tfrac{1}{2}(y_1 + y_7) + y_2 + y_3 + y_4 + y_5 + y_6\right] \\ &= 2\left[\tfrac{1}{2}(36 + 0) + 50 + 56 + 54 + 44 + 26\right] \\ &= \underline{496 \text{ square units}}\end{aligned}$$

(b) By the Mid-ordinate Rule:
The mid-ordinates in this case will be the values of y corresponding to odd values of x

x	1	3	5	7	9	11
y	44	54	56	50	36	14

By the Mid-ordinate Rule:

$$\begin{aligned} \text{area} &= \text{strip width} \times \text{sum of mid-ordinates} \\ &= 2(44 + 54 + 56 + 50 + 36 + 14) \\ &= \underline{512 \text{ square units}} \end{aligned}$$

(c) By Simpson's Rule:
The table of values for even values of x has an odd number of ordinates which is the necessary condition.
By Simpson's Rule:

$$\begin{aligned} \text{area} &= \frac{S}{3}\,[y_1 + y_7 + 4(y_2 + y_4 + y_6) + 2(y_3 + y_5)] \\ &= \frac{2}{3}\,[36 + 0 + 4(50 + 54 + 26) + 2(56 + 44)] \\ &= \underline{504 \text{ square units}} \end{aligned}$$

Note that the figure derived from Simpson's Rule is intermediate between the estimates by the more approximate methods. The figure of 504 sq. units is in fact the correct answer. The error of approximately 1.6% in the other estimates is larger than usual because the boundary of the area in this case is a regular curve which bends one way continually. With a tortuous boundary the small errors would tend to cancel out.

EXERCISE 7

1. The Mid-ordinate Rule, the Trapezoidal Rule, and Simpson's Rule are three formulae used to find irregular areas. State clearly (but do *not* derive) each rule and say which you consider the most accurate of them, giving your reasons.
   On a survey, offsets were taken at intervals along a straight line AB which cuts off the irregular side of a field. The table gives the perpendicular offsets at their respective intervals along AB. Find the area of this irregular portion using the Trapezoidal Rule.

Metres along AB	A 0	20	40	60	70	80	100	120	140	160 B
Offset in metres	0	13	47	50	30	22	19	30	35	0

(E.M.E.U.)

2. A series of parallel measurements (in metres, each 1 m apart) on a small plot of land were 4.5, 16, 28, 40, 53, 64, 69, 66, 63, 60, 56, 52. Use the 'Mid-Ordinate Rule' to find the approximate area of the land in square metres if the first and last measurements represent the boundaries of the plot. (U.L.C.I.)

3. The depth of a horizontal cable below ground varies as follows:

Horizontal distance x (m)	0	40	80	120	160	200	240	280	320
Depth of cable y (m)	2.25	2.5	2.83	3.0	3.08	3.42	3.58	3.17	2.75

   Plot a graph showing variation of depth with length. Using the Mid-ordinate Rule and assuming the trench is 500 mm wide, calculate, correct to the nearest cubic metre, the volume of earth to be dug out to expose the cable throughout its length. (W.J.E.C.)

4. The depth of a stream 20 m wide is measured at different distances from one bank and the results are given in the table:

Distance (m)	1	3	5	7	9	11	13	15	17	19
Depth (m)	1	3.2	3.7	4.7	5	5.5	6	4.7	3	1

Find the area of the cross-section of the stream. (U.L.C.I.)

5. A horizontal trench 1 m wide is dug in uneven ground, and the vertical depth of the trench in mm at 3 m intervals is 215, 200, 180, 175, 160, 125, 105. Find the volume of earth removed. (U.L.C.I.)

6. A plot of land is bounded on two sides by straight roads at right-angles to each other. The other boundary line is irregular and the lengths of perpendicular offsets from one road to the boundary at a common interval of $1\frac{1}{2}$ m are: 14, 16, 15, 15, 14, 15, 14, 14, 13, 12, 10, 12, 12, 13, 12, 10, 8, 7, 7, 6, 6 m. Calculate by Simpson's Rule the area of the plot of land. (U.E.I.)

7. An open channel has a cross-section which is in the form of a rectangle of width which is twice its depth.

The following table shows the depths along a section of the channel taken at 5 m intervals.

Length (m)	0	5	10	15	20	25	30
Depth of channel (m)	10	8	8	7	6	6	5

If the top of the channel is horizontal, what volume of earth would have to be excavated to construct this section of the channel? (N.C.T.E.C.)

8. A water tower is 28 m high and circular in plan. The inside radius of the tower at different heights is given in the following table:

Height (m)	0	7	14	21	28
Radius (m)	10	8	7	6	7

Using Simpson's Rule, find the capacity of the tower in cubic metres. (U.L.C.I.)

9. The areas of horizontal sections of a reservoir at various depths are given in the following table:

Depth of water (m)	0	1.5	2.5	3.5	4.5	6.0	7.5	9.0	10.0
Area of section (m^2)	680	630	540	490	425	260	180	120	0

Using the largest possible convenient scales, plot a smooth curve showing the relationship between area of section and depth. By the use of Simpson's Rule, find from the graph the approximate reduction in the volume of water stored in the reservoir when the depth is decreased from 8 m to 2 m.

10. To find the area of a pond situated near a straight road perpendicular offsets at 10 m intervals were measured to the near and far boundaries of the pond. The results were as follows:

Distance of offsets from one end (m)	0	10	20	30	40	50	60	70	80
Length of offset to near boundary (m)	20	16	4	12	18	23	17	15	20
Length of offset to far boundary (m)	20	30	34	48	58	63	54	32	20

Use Simpson's Rule to find the area of the pond. (U.E.I.)

11. The following table gives the depths at intervals of 1 m of a trench 8 m long and $\frac{1}{2}$ m wide. Calculate the volume, in cubic metres, of earth which had to be taken away to form this trench, assuming an increase in bulk of 25%.

Distance (m)	0	1	2	3	4	5	6	7	8
Depth (mm)	1200	1300	1400	1360	1200	1100	1000	900	800

Use Simpson's Rule to estimate this volume. (E.M.E.U.)

REVISION EXERCISE 7

1. A river is 15 m wide and the average flow of water is 2 m/s. Starting from one bank, the depths at intervals of 1 m are measured and found to be: 1.0, 3.1, 2.9, 2.8, 3.3, 3.8, 3.9, 4.3, 5.2, 5.6, 4.3, 2.1, 1.8, 1.7, 1.2, and 0.8 m in order, from bank to bank. Calculate the flow of water in m^3/s, using the Mid-ordinate Rule to determine the cross-sectional area. (N.C.T.E.C.)

2. A drain is laid between two points A and B 60 m apart, the width of the trench is 1 m. From the following information calculate the volume of earth to be excavated using:
(a) the Trapezoidal Rule; and (b) Simpson's Rule.
The levels are measured from a common datum.

Distance	0	10	20	30	40	50	60
Ground level	53.4	53.0	52.1	51.6	52.7	52.9	53.3
Level of trench bottom	50.1	49.6	49.1	48.6	48.1	47.6	47.1

(N.C.T.E.C.)

3. During the construction of a reservoir to supply a new town, it was noted that the water in the river which was to be used to fill it flowed at 5 m/s. The river was 10 m wide and the following soundings were taken at intervals of 1 m from the bank:

Distance from bank (m)	0	1	2	3	4	5	6	7	8	9	10
Depth (m)	0	2	3	5	7	10	10	8	7	6	0

Calculate the daily supply of water in metric tons. (1 tonne = 1000 kg) (E.M.E.U.)

4. The following table gives the lengths of the offsets from the edge of a straight road to the boundary of a building site.

Distance (m)	0	12	24	36	48	60	72	84	96	108	120
Offset (m)	35	30	32	29	40	36	41	49	56	62	69

Using Simpson's Rule, calculate the approximate area of the site. (W.J.E.C.)

5. A trench 10 m long and 1 m wide is dug in undulating ground to form a level foundation. The depths of the trench at the stated distances from one end are as follows:

Distance (m)	0	1	2	3	4	5	6	7	8	9	10
Depth (m)	7.0	6.5	5.5	4.75	5.0	4.5	3.0	3.5	4.0	4.25	5.0

Calculate, correct to the nearest cubic metre, the volume of excavation in the trench. (U.E.I.)

6. For the graph of $x^2+4y^2 = 100$, calculate the values of y which correspond to the following values of x:

$$-10, -8, -6, -4, -2, 0, 2, 4, 6, 8, 10.$$

Use Simpson's Rule to find the area of this ellipse and use the formula $A = \pi ab$ to check your result.

7. An open channel has a cross-section which is in the form of a trapezium whose horizontal base width is 3 m and the side slopes are inclined at 60° to the horizontal.

The following table shows the depths along a section of the channel taken at 5 m intervals.

Length (m)	0	5	10	15	20	25	30
Depth of channel (m)	6	4	4	3	2	2	1

What volume of earth would have to be excavated to construct this section of the channel? (N.C.T.E.C.)

8. The widths of a boating lake at varying distances from one end are given in the following table:

Distance (m)	0	6	12	20	34	42	52	64	68	76	80
Width (m)	3.0	8.2	12.1	14.2	13.8	13.0	12.5	12.1	9.3	4.2	3.4

Plot a graph of width against distance and calculate the surface area of the lake by Simpson's Rule, using 8 intervals. Take 10 mm for 4 m as distance scale and 10 mm for 1 m as width scale. (U.E.I.)

9. For a uniformly loaded cantilever of length l, it is known that the deflection in m, at a distance x_1 from the fixed end is given by the area from $x = 0$ to $x = x_1$ under the graph

$$\frac{w}{6EI}[l^3 - (l - x)^3]$$

against x, where w, E, I, l are constants.

Tabulate values of the expression $l^3 - (l - x)^3$ for values of x from 0 to 12 at intervals of 2 m for the case $l = 12$ m, and use Simpson's Rule to determine the deflection at the end of a cantilever of length 12 m for which $w = 54$, $E = 7 \times 10^9$, $I = 0.001$. (D.D.C.T.)

8 Solution of Triangles

A RIGHT-ANGLED triangle may always be solved by the straightforward application of sine, cosine or tangent together with the theorem of Pythagoras. For a triangle which is not right-angled, we may use the sine rule, the cosine rule or the tangent rule together with the fact that the sum of the angles of any triangle is 180°.

The sine rule

In chapter 6 we showed that the area of a triangle is given by any of the three forms:

$$\tfrac{1}{2}bc \sin A = \tfrac{1}{2}ac \sin B = \tfrac{1}{2}ab \sin C$$

From this, if we divide through by $\tfrac{1}{2}abc$:

$$\frac{\sin A}{a} = \frac{\sin B}{b} = \frac{\sin C}{c}$$

This relationship is known as the *sine rule*, and can be used either in the above form or in the inverted form:

$$\frac{a}{\sin A} = \frac{b}{\sin B} = \frac{c}{\sin C}$$

Alternatively, the sine rule may be derived as follows:

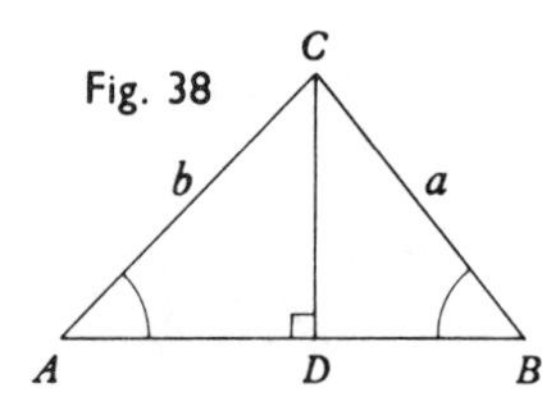

From triangle ACD (fig. 38), $CD = b \sin A$.
From triangle BCD (fig. 38), $CD = a \sin B$.

$$\therefore \quad a \sin B = b \sin A$$

i.e.

$$\frac{a}{\sin A} = \frac{b}{\sin B}$$

By dropping a perpendicular from A on to BC and applying the same method,

$$\frac{b}{\sin B} = \frac{c}{\sin C}$$

and thus,

$$\frac{a}{\sin A} = \frac{b}{\sin B} = \frac{c}{\sin C}$$

The sine rule may be used, when given two sides and an angle opposite to one of them, to find another angle, or, when given one side and two angles, to find another side.

Example 1 In a triangle ABC, $a = 4$, $b = 5$ and angle $A = 52°$. Find angle B.

From the sine rule:

$$\frac{a}{\sin A} = \frac{b}{\sin B}$$

i.e.

$$\frac{4}{\sin 52°} = \frac{5}{\sin B}$$

$$\sin B = \frac{5 \sin 52°}{4}$$

$$\sin B = \frac{5 \times 0.7880}{4}$$

$$\sin B = 0.9850$$

$$\therefore \quad B = 80° \, 4'$$

Example 2 A workshop 8 m wide has a span roof which slopes at 35° on one side and 42° on the other. Find the length of the roof slopes.

Figure 39 shows a section of the roof. By subtraction the angle at the ridge must be 103°.

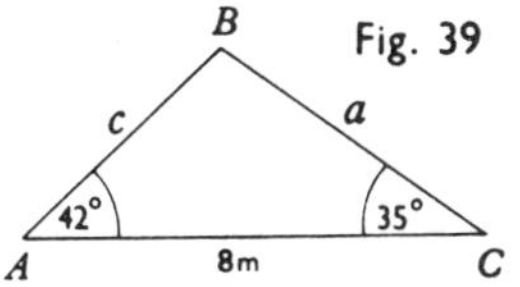

Fig. 39

Applying the sine rule:

$$\frac{a}{\sin 42°} = \frac{8}{\sin 103°}$$

$$a = \frac{8 \sin 42°}{\sin 103°}$$

$$= 5.49 \text{ m.}$$

Similarly:

$$\frac{c}{\sin 35°} = \frac{8}{\sin 103°}$$

$$c = \frac{8 \sin 35°}{\sin 103°}$$

$$= 4.60 \text{ m.}$$

i.e. the two roof slopes are 5.49 m and 4.60 m (correct to the nearest 10 mm).

No.	*Log*
8	0.9031
sin 42°	$\bar{1}.8255$
	0.7286
sin 77°	$\bar{1}.9887$
5.494	0.7399
8	0.9031
sin 35°	$\bar{1}.7586$
	0.6617
sin 77°	$\bar{1}.9887$
4.603	0.6730

The cosine rule

In fig. 40 BD is the perpendicular from B on to AC. $AD = x$, and therefore $DC = b - x$.

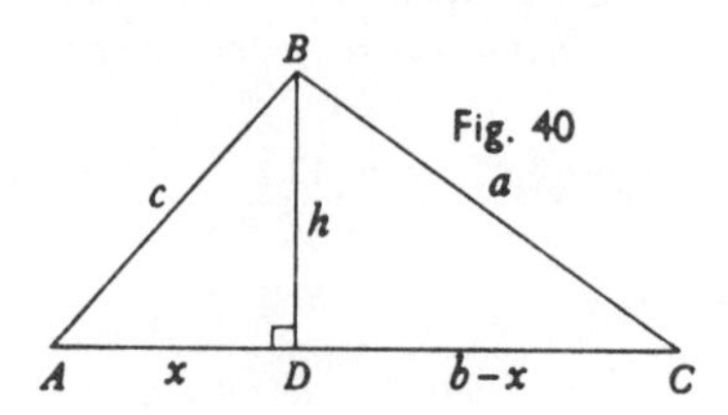

Applying the theorem of Pythagoras to each of the triangles ABD and BCD

$$a^2 = h^2 + (b - x)^2$$

i.e. $$a^2 = h^2 + b^2 - 2bx + x^2$$

and $$c^2 = h^2 + x^2$$

subtracting, $$a^2 - c^2 = b^2 - 2bx$$

but $$x = c \cos A$$

therefore $$a^2 - c^2 = b^2 - 2b.c \cos A$$

i.e. $$a^2 = b^2 + c^2 - 2bc.\cos A$$

Similarly, $$b^2 = a^2 + c^2 - 2ac.\cos B$$

and $$c^2 = a^2 + b^2 - 2ab.\cos C$$

These forms are used to find the third side of a triangle when given two sides and the angle between them.

If all three sides are known, and it is required to find an angle, then the cosine rule is more conveniently expressed in one of the forms

$$\cos A = \frac{b^2 + c^2 - a^2}{2bc}$$

$$\cos B = \frac{a^2 + c^2 - b^2}{2ac}$$

$$\cos C = \frac{a^2 + b^2 - c^2}{2ab}$$

Example 1 Because of obstructions it is not possible to measure directly the distance between two points A and B. Measurements are therefore taken from another point C as follows: $AC = 66$ m, $BC = 40$ m, angle $ACB = 30°$. From these measurements calculate the distance AB.

See fig. 41.

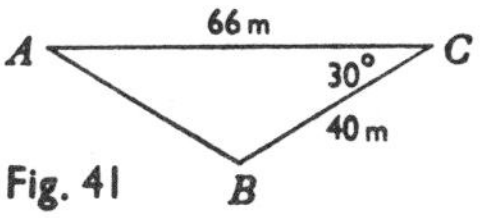

Fig. 41 (triangle ABC with $AC = 66$ m, $BC = 40$ m, angle $C = 30°$)

$$c^2 = a^2 + b^2 - 2ab.\cos C$$
$$c^2 = 40^2 + 66^2 - 2.40.66.\cos 30°$$
$$= 1600 + 4356 - 5280 \times 0.8660$$
$$= 5956 - 4572.48$$
$$= 13\,832.5$$
$$\therefore\ c = 37.2 \text{ m}$$

Example 2 Find the largest angle in a triangle with sides 3 m, 5 m and 7 m.

The largest angle must be opposite the largest side (see fig. 42).

Fig 42 (triangle ABC with $AB = 3$, $AC = 5$, $BC = 7$)

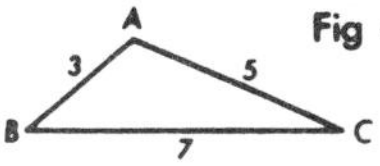

$$\cos A = \frac{b^2 + c^2 - a^2}{2bc}$$
$$= \frac{25 + 9 - 49}{2.3.5} = \frac{-15}{30} = -\tfrac{1}{2}$$
$$\therefore\ A = 120°$$

The tangent rule

The tangent rule may be derived from the sine rule as follows:

Let
$$\frac{a}{\sin A} = \frac{b}{\sin B} = k$$

from which
$$a = k \sin A \text{ and } b = k \sin B$$

then
$$\frac{a - b}{a + b} = \frac{k \sin A - k \sin B}{k \sin A + k \sin B}$$
$$= \frac{\sin A - \sin B}{\sin A + \sin B}$$
$$= \frac{2 \cos \frac{1}{2}(A + B).\sin \frac{1}{2}(A - B)}{2 \sin \frac{1}{2}(A + B).\cos \frac{1}{2}(A - B)}$$
$$\therefore\ \frac{a - b}{a + b} = \frac{\tan \frac{1}{2}(A - B)}{\tan \frac{1}{2}(A + B)}$$

or, since
$$A + B = 180° - C$$

so
$$\tfrac{1}{2}(A + B) = 90° - \frac{C}{2}$$
$$\therefore\ \tan \tfrac{1}{2}(A + B) = \cot \frac{C}{2}$$

and substitution of this gives the alternative form of the tangent rule
$$\tan \frac{A - B}{2} = \frac{a - b}{a + b}\cdot\cot \frac{C}{2}$$

similarly $$\tan\frac{A - C}{2} = \frac{a - c}{a + c}\cdot\cot\frac{B}{2}$$

and $$\tan\frac{B - C}{2} = \frac{b - c}{b + c}\cdot\cot\frac{A}{2}$$

Geometrical proof

In fig. 43 ABC is the triangle. With C as centre, and CA as radius describe a circle to cut BC at D and BC produced at E; then $BD = a - b$ and $BE = a + b$ and angle DAE is a right-angle. DF is drawn at right-angles to DA to meet AB in F.

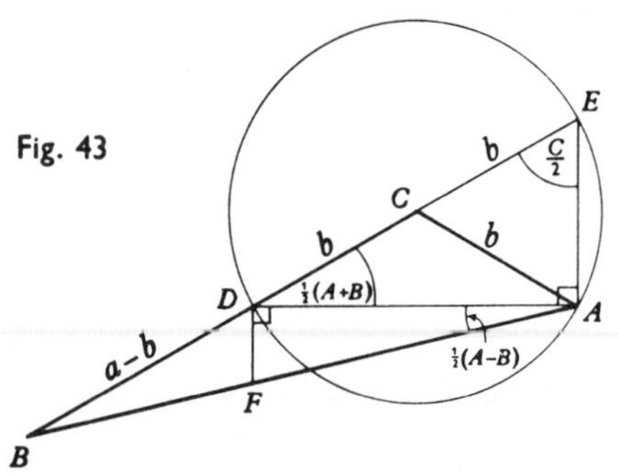

Fig. 43

ACE is an isosceles triangle and thus angle $AEC = C/2$,

therefore angle $ADE = 90° - C/2 = \frac{1}{2}(A + B)$

Also angle $BDF = C/2$ so triangles BDF and BEA are similar.

ACD is an isosceles triangle and thus angle CAD is also $\frac{1}{2}(A + B)$, but angle CAD + angle BAD = angle A, therefore angle $BAD = \frac{1}{2}(A - B)$

$$\frac{\tan\frac{1}{2}(A - B)}{\tan\frac{1}{2}(A + B)} = \frac{\tan BAD}{\tan ADE} = \frac{DF/AD}{EA/AD} = \frac{DF}{EA} = \frac{BD}{BE} = \frac{a - b}{a + b}$$

Hence $$\tan\frac{A - B}{2} = \frac{a - b}{a + b}\cot\frac{C}{2}$$

Use of the tangent rule

Given two sides and the included angle of any triangle, the tangent rule may be used to find the remaining angles.

Example: Given a triangle ABC in which $a = 10$, $b = 6$, $C = 120°$, find A and B.

Using the tangent rule $$\tan\frac{A - B}{2} = \frac{10 - 6}{10 + 6}\cot 60° = 0.144\,34$$

$$\therefore\quad \tfrac{1}{2}(A - B) = 8°\,13'$$

but $$\tfrac{1}{2}(A + B) = 30°$$

hence $$A = 38°\,13' \text{ and } B = 21°\,47'$$

EXERCISE 8(a)

1. In a triangle the smallest angle is 40°, and one other angle is 57°. The shortest side is 1 metre. Calculate the lengths of the other two sides. (U.L.C.I.)

2. In a triangle ABC the side $BC = 3.2$ m and angles B and C are 47° and 35° respectively. Calculate the length of AC. (U.L.C.I.)

3. A jib crane consists of a vertical post $PQ = 4$ m, the inclined jib $QS = 8$ m, and PS is the tie. Angle $QPS = 125°$. Calculate (a) the inclination of QS to the vertical; (b) the length of PS. (N.C.T.E.C.)

4. A roof truss for a couple roof of unequal slopes is in the form of a triangle ABC. If the tie bar AC is 8 m long, the principal rafter AB is 6 m long and angle CAB is 38°, find (a) BC, (b) angle ABC. (U.L.C.I.)

5. Using the formula: $$\tan\frac{B-C}{2} = \frac{b-c}{b+c}\cot\frac{A}{2}$$ calculate the angles B and C of the triangle ABC given that $c = 30$, $b = 36$, angle $A = 57°$. Also calculate the length of BC. (E.M.E.U.)

6. In a traverse survey, it is not possible to measure directly the distance between two points B and C and measurements are made to a third point A, as follows: $AB = 512$ m, $AC = 380$ m. The angle BAC is observed to be 108°. Calculate the distance BC and the angles ABC and ACB. (U.L.C.I.)

7. D and C are two inaccessible points and are viewed from observation points A and B where the horizontal distance $AB = 70$ m. D and C are on the same side of AB. The horizontal angles measured are:

   $DAB = 104°\ 10'$ $CBA = 81°\ 37'$ $CAB = 52°\ 25'$ $DBA = 29°\ 16'$

   Calculate the horizontal distance DC. (N.C.T.E.C.)

The 's' formulae

Starting from the cosine rule:

$$\cos A = \frac{b^2 + c^2 - a^2}{2bc}$$

$$1 - \cos A = 1 - \frac{b^2 + c^2 - a^2}{2bc} \qquad\Big|\qquad 1 + \cos A = 1 + \frac{b^2 + c^2 - a^2}{2bc}$$

$$2\sin^2\frac{A}{2} = \frac{2bc - b^2 - c^2 + a^2}{2bc} \qquad\Big|\qquad 2\cos^2\frac{A}{2} = \frac{2bc + b^2 + c^2 - a^2}{2bc}$$

$$\sin^2\frac{A}{2} = \frac{a^2 - (b-c)^2}{4bc} \qquad\Big|\qquad \cos^2\frac{A}{2} = \frac{(b+c)^2 - a^2}{4bc}$$

factorising as the difference of two squares,

$$\sin^2\frac{A}{2} = \frac{(a-b+c)(a+b-c)}{4bc} \qquad\Big|\qquad \cos^2\frac{A}{2} = \frac{(b+c+a)(b+c-a)}{4bc}$$

writing s for $\frac{1}{2}(a + b + c)$

$$\sin^2\frac{A}{2} = \frac{(2s-2b)(2s-2c)}{4bc} \qquad\Big|\qquad \cos^2\frac{A}{2} = \frac{2s(2s-2a)}{4bc}$$

$$\therefore\quad \sin\frac{A}{2} = \sqrt{\frac{(s-b)(s-c)}{bc}} \qquad\Big|\qquad \cos\frac{A}{2} = \sqrt{\frac{s(s-a)}{bc}}$$

by division,
$$\tan\frac{A}{2} = \sqrt{\frac{(s-b)(s-c)}{s(s-a)}}$$

These formulae are used with logarithms. If a single angle is required the cosine form is the quickest, but if all angles are required the tangent form should be used.

Also:
$$\sin A = 2\sin\frac{A}{2}\cos\frac{A}{2}$$

and thus the area of a triangle $= \frac{1}{2}bc\sin A = bc\sin\frac{A}{2}\cos\frac{A}{2}$

$$= bc\sqrt{\frac{(s-b)(s-c)}{bc}}\cdot\sqrt{\frac{s(s-a)}{bc}}$$

$$= \sqrt{s(s-a)(s-b)(s-c)}$$

Example: Find the three angles and the area of a triangle with sides 13 m, 14 m and 15 m.

$$a = 13\text{ m} \qquad s = 21\text{ m} \qquad s-a = 8\text{ m}$$
$$b = 14\text{ m} \qquad s-b = 7\text{ m}$$
$$c = 15\text{ m} \qquad s-c = 6\text{ m}$$
$$2s = 42\text{ m}$$

$$\tan\frac{A}{2} = \sqrt{\frac{(s-b)(s-c)}{s(s-a)}} = \sqrt{\frac{7\times 6}{21\times 8}} = \sqrt{\frac{1}{4}} = \frac{1}{2}$$

$$\tan\frac{B}{2} = \sqrt{\frac{(s-c)(s-a)}{s(s-b)}} = \sqrt{\frac{6\times 8}{21\times 7}} \quad \sqrt{\frac{16}{49}} = \frac{4}{7}$$

$$\tan\frac{C}{2} = \sqrt{\frac{(s-a)(s-b)}{s(s-c)}} = \sqrt{\frac{8\times 7}{21\times 6}} \quad \sqrt{\frac{4}{9}} = \frac{2}{3}$$

$$\therefore \quad \frac{A}{2} = 26^\circ\,34' \qquad A = 53^\circ\,8'$$

and
$$\frac{B}{2} = 29^\circ\,44\tfrac{1}{2}' \qquad B = 59^\circ\,29'$$

and
$$\frac{C}{2} = 33^\circ\,41\tfrac{1}{2}' \qquad C = 67^\circ\,23'$$

Check total $= 180^\circ\,00'$

$$\text{area} = \sqrt{s(s-a)(s-b)(s-c)} = \sqrt{21.8.7.6} = \sqrt{7056} = 84\text{ m}^2$$

EXERCISE 8(b)

1. Calculate the value of the angle A of a triangle ABC from the formula
$$\tan\frac{A}{2} = \sqrt{\frac{(s-b)(s-c)}{s(s-a)}}$$
where $2s = a + b + c$; and $a = 13.6$ m, $b = 8.45$ m and $c = 7.65$ m. (N.C.T.E.C.)

2. Using the formula:
$$\cos\frac{A}{2} = \sqrt{\frac{s(s-a)}{bc}}$$
find the largest angle of a triangle with sides 6 m, 9 m and 11 m. Find also the area of this triangle.

3. Find the angles and the area of a triangle with sides 10 m, 17 m and 21 m.

4. Given that $AB = 16$ m, $BC = 12$ m, $CD = 21$ m, $DA = 29$ m and $AC = 20$ m, find the area of the plot of land $ABCD$. Find also the angles A, B, C and D.

5. A plot of land $ABCD$ is in the form of a parallelogram. $AB = 205$ m, $AC = 223$ m and $AD = 46$ m. Find the area of this plot of land. Find also the length of BD.

6. In triangle ABC, $AB = 20$ m, $BC = 21$ m and $CA = 13$ m. Use the s formula to find the area of this triangle, and hence find the length of the perpendicular from A on to BC.

REVISION EXERCISE 8

1. The area of a triangle ABC is 40 m², the angle B is 30°, and the side BC is 8 m. Solve the triangle. (U.L.C.I.)

2. Prove that in the acute angled triangle ABC:
$$\frac{a}{\sin A} = \frac{b}{\sin B} = \frac{c}{\sin C}$$
If $c = 3.5$ m, $b = 2.8$ m and angle $A = 47°$, find a and the angles B and C. (U.L.C.I.)

3. A survey of a field, triangular in shape, gave the following results:
$AB = 300$ m $\quad BC = 240$ m $\quad$ angle $ABC = 60°$
Calculate the area of the field in square metres. Find also the length of AC. (K.C.E.A.B.)

4. A road runs for 2 km in a direction 067° from a point A to a point B, and then for $1\frac{1}{2}$ km in a direction 332° to a point C. Find the distance and bearing of C from A. (U.E.I.)

5. A, B, C and D are stations in a survey forming a quadrilateral such that $AB = 148.5$ m, $AD = 126.6$ m, angle $ABC = 70° 33'$, angle $BAC = 41° 19'$, and angle $BAD = 79° 41'$. Calculate CD. (W.J.E.C.)

6. The plan of a building plot is a quadrilateral $ABCD$, in which angle $DAB = 80°$, $AB = 50$ m, $BC = 60$ m, $CD = 32$ m, and diagonal $BD = 66$ m. Calculate the angle ABC and the length of the diagonal AC. (U.L.C.I.)

7. Prove the area of a triangle is half the product of two sides and the sine of the included angle.

   Two sides of a triangular plot of ground are known to be 30 m and 24 m respectively. Assuming the area, stated as 300 m² is correct, find the length of the third side and the magnitude of the angles of the triangle. (U.L.C.I.)

8. A quadrilateral, *ABCD*, is the plan of the lines of a survey. Points *A* and *B* are 100 m apart and angles observed at these points are $BAC = 64°$, $CAD = 66°$, $ABD = 38°$, $DBC = 48°$. Calculate the distance *CD*. (U.L.C.I.)

9. *ABCD* is a plot of ground. $AB = BC = CD = 100$ m. $DA = 50$ m and angle $DAB = 110°$. Find the length of *BD* and the size of angles *B*, *C* and *D*. (E.M.E.U.)

10. Using the formula

$$\tan\frac{A}{2} = \sqrt{\frac{(s-b)(s-c)}{s(s-a)}}$$

find the greatest angle of a triangle having sides 7.3 m, 6.8 m and 9.7 m. (E.M.E.U.)

11. Part of a jib crane is represented by a triangle *ABC* in which *AB* is the vertical post 4.0 m high, the point *A* being on the ground. *BC* is the tie rope and *AC* is the jib, which is 8.2 m long. When the angle *BAC* is 38°, calculate:
    (a) the vertical height of *C* above the ground;
    (b) the length of the rope *BC*. (N.C.T.E.C.)

12. In a survey of a field the survey lines were in the shape of a quadrilateral *ABCD*. The lengths of the sides were $AB = 78$ m, and $BC = 90$ m, $CD = 70$ m and $DA = 113$ m. The internal angle at *A* was 75°. Determine the lengths of the two diagonals *AC* and *BD*. (N.C.T.E.C.)

13. The distance between two poles *A* and *B* on a level building site is 100 m. At a fixed point on the site there is a third pole *C*. By means of a theodolite two readings from the base line *AB* were taken. It was found that angle $CAB = 70°$ and angle $CBA = 35°$. Find:
    (a) the length of *AC* and *BC*;
    (b) the area contained in the triangle *ABC*. (K.C.E.A.B.)

14. Figure 44 shows the main lines of a survey which forms a triangle *ABC*. Find the lengths of *AC* and *AB*. (U.L.C.I.)

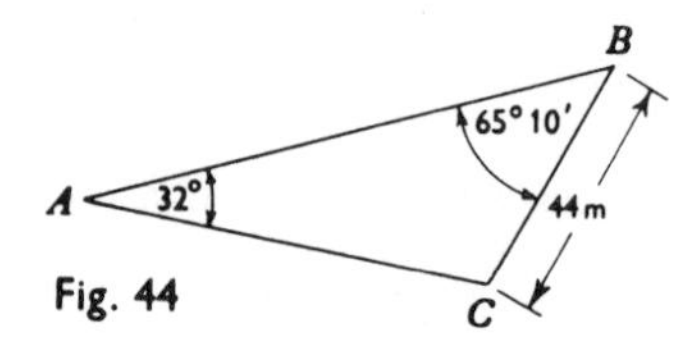

Fig. 44

9 Calculus

Gradient of chord and tangent

Consider the curve $y = x^2$ a portion of which is shown in fig. 45. Let P be the point at which $x = 1$ and $y = 1$ [which may be abbreviated to (1, 1), writing the x-value first]. Q is another point on the curve and we shall consider it to start at the point (5, 25) and to move down the curve towards P.

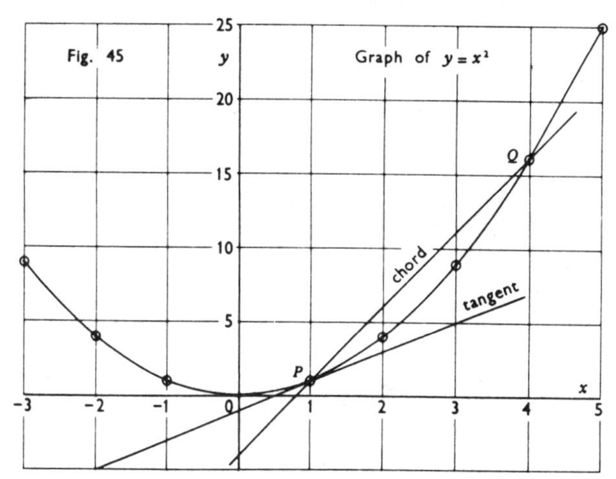

When Q is at (5, 25) the gradient of chord PQ is $\dfrac{25 - 1}{5 - 1} = \dfrac{24}{4} = 6$

When Q is at (4, 16) the gradient of chord PQ is $\dfrac{16 - 1}{4 - 1} = \dfrac{15}{3} = 5$

When Q is at (3, 9) the gradient of chord PQ is $\dfrac{9 - 1}{3 - 1} = \dfrac{8}{2} = 4$

When Q is at (2, 4) the gradient of chord PQ is $\dfrac{4 - 1}{2 - 1} = \dfrac{3}{1} = 3$

Then taking a point an equivalent distance on the other side of P.

When Q is at (0, 0) the gradient of chord PQ is $\dfrac{1 - 0}{1 - 0} = 1$

The gradients of the chords thus form a series:

$$6, 5, 4, 3, -, 1$$

The missing gradient is that for $x = 1$, but when Q reaches this point it coincides with point P. If the chord PQ is regarded as being part of a longer straight line which simply intersects the curve at the two points P and Q, then, when Q coincides with P, the straight line becomes a tangent to the curve at the point where P and Q coincide. It would therefore appear reasonable from the series which we have obtained to assume that the gradient of the tangent at P would be 2. Let us now look at this a little more closely by considering values for Q close to P.

When Q is at (1.5, 2.25) the gradient of chord PQ is $\frac{1.25}{0.5} = 2.5$

When Q is at (1.4, 1.96) the gradient of chord PQ is $\frac{0.96}{0.4} = 2.4$

When Q is at (1.3, 1.69) the gradient of chord PQ is $\frac{0.69}{0.3} = 2.3$

When Q is at (1.2, 1.44) the gradient of chord PQ is $\frac{0.44}{0.2} = 2.2$

When Q is at (1.1, 1.21) the gradient of chord PQ is $\frac{0.21}{0.1} = 2.1$

When Q is at (1.01, 1.0201) the gradient of chord PQ is $\frac{0.0201}{0.01} = 2.01$

When Q is at (1.001, 1.002 001) the gradient of chord PQ is $\frac{0.002\,001}{0.001} = 2.001$

It thus becomes increasingly obvious that the closer Q approaches to P, the nearer the gradient is to 2.

Now instead of selecting some special values, let us take a general case with co-ordinates for P (x, y), and Q a point very close to it on the curve $y = x^2$. Let the x co-ordinate for Q be $x + \delta x$, where δx is a small increase in x. (δ is the Greek letter ' delta '.)

Then,
$$y = x^2$$
letting x increase by a small amount δx, results in a corresponding smal change δy in y.

Thus
$$y + \delta y = (x + \delta x)^2$$
$$y + \delta y = x^2 + 2x\delta x + (\delta x)^2$$

Subtracting the original equation:

$$\delta y = 2x\delta x + (\delta x)^2$$

Dividing by δx to obtain the gradient of the chord:

$$\frac{\delta y}{\delta x} = 2x + \delta x$$

Now as Q approaches P, δx tends to zero, and the ratio $\delta y/\delta x$, which is the gradient of the chord, tends in the limit to the gradient of the tangent, which we denote by the symbol dy/dx.

Thus,
$$\underset{\delta x \to 0}{\text{Limit}} \left(\frac{\delta y}{\delta x}\right) = \frac{dy}{dx} = 2x$$

Thus when $x = 1$, the gradient of the tangent is 2 exactly.

The process of finding the value of dy/dx is called **differentiation** and when each stage of the above procedure is to be set down it is called **differentiation from first principles.** The symbol dy/dx stands for 'the differential coefficient of y with respect to x'.

Example 1 Differentiate from first principles $y = \dfrac{1}{x}$.

Let x increase by a small amount δx.
Let the corresponding increase in y be δy.

$$y + \delta y = \frac{1}{x + \delta x}$$

Subtracting the original equation:

$$\delta y = \frac{1}{x + \delta x} - \frac{1}{x} = \frac{x - (x + \delta x)}{x(x + \delta x)} = \frac{-\delta x}{x(x + \delta x)}$$

Divide by δx:

$$\frac{\delta y}{\delta x} = \frac{-1}{x(x + \delta x)}$$

In the limit as $\delta x \to 0$, $\dfrac{\delta y}{\delta x} \to \dfrac{dy}{dx}$

$$\therefore \quad \frac{dy}{dx} = -\frac{1}{x^2}$$

Example 2 Differentiate from first principles $y = x^n$.

Let x increase by a small amount δx.
Let the corresponding increase in y be δy.

$$y + \delta y = (x + \delta x)^n$$

Expanding the right-hand side by the binomial theorem:

$$y + \delta y = x^n + n\,x^{n-1}\delta x + \tfrac{1}{2}n(n - 1)\,x^{n-2}(\delta x)^2 + \ldots$$

Subtracting the original equation:

$$\delta y = n\,x^{n-1}\,\delta x + \tfrac{1}{2}n(n - 1)x^{n-1}(\delta x)^2 + \ldots$$

Divide by δx:

$$\frac{\delta y}{\delta x} = n\,x^{n-1} + \tfrac{1}{2}n(n-1)x^{n-1}\delta x + \text{other terms in } \delta x.$$

In the limit as $\delta x \to 0$, $\dfrac{\delta y}{\delta x} \to \dfrac{dy}{dx}$,

$$\therefore \quad \frac{dy}{dx} = n\,x^{n-1}$$

The above result is true for *all* values of n, including negative and fractional values, and its use as a formula enables us to short-cut the process of differentiation from first principles in many cases.

Examples:

(a) $y = x^4$ $\qquad \dfrac{dy}{dx} = 4x^3$

(b) $y = \dfrac{1}{x^3} = x^{-3}$ $\qquad \dfrac{dy}{dx} = -3x^{-4} = -\dfrac{3}{x^4}$

(c) $y = \sqrt{x^3} = x^{\frac{3}{2}}$ $\qquad \dfrac{dy}{dx} = \dfrac{3}{2}x^{\frac{1}{2}} = \dfrac{3\sqrt{x}}{2}$

(d) $y = \dfrac{1}{\sqrt{x}} = x^{-\frac{1}{2}}$ $\qquad \dfrac{dy}{dx} = -\dfrac{1}{2}x^{-\frac{3}{2}} = -\dfrac{1}{2\sqrt{x^3}}$

(e) $y = 1 = x^0$ $\qquad \dfrac{dy}{dx} = 0$

Differentiation of constants

When a power of x is multiplied by a constant, that constant remains unchanged by the process of differentiation:

i.e., if $y = a\,x^n$, $\qquad \dfrac{dy}{dx} = a\,nx^{n-1}$

The product an will then be combined into a single new constant.

If any isolated constant is differentiated, the result will be zero.

Examples:

(a) $y = 5x^{1.2}$ $\qquad \dfrac{dy}{dx} = 5 \times 1.2x^{0.2} = 6x^{0.2}$

(b) $y = \dfrac{3\sqrt{x}}{2} = \dfrac{3}{2}x^{\frac{1}{2}}$ $\qquad \dfrac{dy}{dx} = \dfrac{3}{2} \times \dfrac{1}{2}x^{-\frac{1}{2}} = \dfrac{3}{4\sqrt{x}}$

(c) $y = 2(a + b)$ $\qquad \dfrac{dy}{dx} = 0$

Differentiation of a sum of several terms

The rule is to differentiate term by term and combine the results.

Examples:

(a) $y = ax^2 + bx + c$ $\qquad \dfrac{dy}{dx} = 2ax + b$

(b) $y = 4x^3 - 2 - \dfrac{4}{x^3}$ $\qquad \dfrac{dy}{dx} = 12x^2 + \dfrac{12}{x^4}$

(c) $y = (3x - 2)^4 = 81x^4 - 216x^3 + 216x^2 - 96x + 16$

$$\frac{dy}{dx} = 324x^3 - 628x^2 + 432x - 96$$

$$= 12(3x - 2)^3$$

Differentiation of a function of a function

It is obvious that in the last of these examples, the power to which the whole bracket was raised was reduced by one in differentiation. Let us now examine this to see whether it is possible to differentiate such *without* multiplying out the bracket.

$$y = (3x - 2)^4$$

put $\qquad X = 3x - 2$

then $\qquad y = X^4$

and $\qquad \dfrac{dy}{dX} = 4X^3 = 4(3x - 2)^3$

Differentiating $\qquad X = 3x - 2$

we get $\qquad \dfrac{dX}{dx} = 3$

We have already shown that the result of differentiating $(3x - 2)^4$ is $12(3x - 2)^3$, and it can now be seen that this is the product of $\dfrac{dy}{dX}$ and $\dfrac{dX}{dx}$

i.e. $\qquad \dfrac{dy}{dx} = \dfrac{dy}{dX} \cdot \dfrac{dX}{dx}$

We thus deduce the general rule, that if

$$y = (ax + b)^n \qquad \frac{dy}{dx} = an(ax + b)^{n-1}$$

The full proof of this is beyond the scope of this book, but is true for *all* values of n, including negative and fractional values.

Examples:

(a) $y = 2(5x + 1)^3$ $\qquad \dfrac{dy}{dx} = 30(5x + 1)^2$

(b) $y = (1 - 7x)^5$ $\qquad \dfrac{dy}{dx} = -35(1 - 7x)^4$

(c) $y = \sqrt{(8x - 3)}$ $\qquad \dfrac{dy}{dx} = 8 \times \frac{1}{2}(8x - 3)^{-\frac{1}{2}} = \dfrac{4}{\sqrt{(8x - 3)}}$

The above process may be extended to include other powers of x within the bracket, the rule being to regard the bracket as an entity raised to the given power in the first place, and subsequently to differentiate the content of the bracket and multiply the results together.

Examples:

(a) $y = (x^2 + 3)^5$ $\qquad \dfrac{dy}{dx} = 5(x^2 + 3)^4 \times 2x = 10x(x^2 + 3)^4$

(b) $y = (ax^2 + bx + c)^n$ $\qquad \dfrac{dy}{dx} = n(ax^2 + bx + c)^{n-1} \times (2ax + b)$

$$= n(2ax + b)(ax^2 + bx + c)^{n-1}$$

(c) $y = (ax^m + b)^n$ $\qquad \dfrac{dy}{dx} = n(ax^m + b)^{n-1} \times amx^{m-1}$

$$= amnx^{m-1}(ax^m + b)^{n-1}$$

(d) $y = \sqrt{(5 - 3x^4)}$ $\qquad \dfrac{dy}{dx} = \frac{1}{2}(5 - 3x^4)^{-\frac{1}{2}} \times (-12x^3)$

$$= -\frac{6x^3}{\sqrt{(5 - 3x^4)}}$$

EXERCISE 9(a)

1. Differentiate $y = x^3$ from first principles.

2. If P, Q and R are the points (1, 1), (2, 8) and (3, 27) respectively on the graph of $y = x^3$, find the gradients of the tangents at the points P, Q and R, and the gradient of the chord PR.

3. Write down the differential coefficients of the following:
   (a) x^5 (b) $x^{3.5}$ (c) $x^{\frac{3}{2}}$ (d) $7x^{-4}$ (e) $4x^{\frac{5}{4}}$
   (f) $2x^{-\frac{3}{4}}$ (g) $\frac{1}{3}x^{-6}$ (h) $\frac{7}{8}x^{\frac{1}{3}}$ (i) $9x^{-0.02}$ (j) $x^{1/n}$

4. Differentiate the following:
   (a) $y = \dfrac{1}{x^9}$ (b) $y = \dfrac{1}{\sqrt{x}}$ (c) $y = \dfrac{1}{3x}$
   (d) $y = \dfrac{1}{5x^4}$ (e) $y = 2\sqrt{x}$ (f) $y = \sqrt{2x}$

5. Differentiate:

(a) $2x^3 + 3x^2 - 4x + 5$ (b) $2x\sqrt{x} - 4\sqrt{x}$

6. Differentiate:

(a) $(5x - 3)^2$ (b) $(2 - x)^3$ (c) $\sqrt{(4 - 3x)}$

(d) $\dfrac{1}{5 - x}$ (e) $\dfrac{7}{(4x - 3)^2}$ (f) $\dfrac{4}{\sqrt{(1 - 3x)}}$

7. Differentiate $y = 2\sqrt{ax}$. Find the gradient of the tangents to the curve at the points where $x = 1$, a, and $4a$.
At what point on the curve is the gradient $2a$?

8. Find $\dfrac{dy}{dx}$ in the following cases:

(a) $y = (x^3 - 4)^2$ (b) $y = (1 + x + x^2)^3$ (c) $y = \dfrac{1}{(6x^2 + 5)^4}$

(d) $y = (2 + \sqrt{x})^3$ (e) $y = \sqrt{(x^2 - 2x)}$

Choice of variable

Before an equation can be differentiated in the way shown above, it is necessary for the equation to be arranged so that there is one variable on each side of the equals sign. The symbols used need not be x and y, but it is conventional to use letters from the second half of the alphabet to represent variables. In the equation $s = 20t - 4.9t^2$ we have t as the independent variable (in place of x) and s as the dependent variable (in place of y). If this equation had been given as $s = ut + \frac{1}{2}at^2$ it would not have been so obvious which letters represented the variables. Given that u represented an initial velocity of 20 m/s upwards and that a was the acceleration of 9.8 m/s² downwards, we could substitute the values of these constants and thus obtain the first equation. Having established which symbols represent the variables, it is then possible to differentiate.

$$s = 20t - 4.9t^2$$

$$\frac{ds}{dt} = 20 - 9.8t$$

Similarly, with an equation such as $A = \frac{1}{2}r^2\theta$ for the area of a sector of a circle, it is necessary to determine from the problem which of the quantities A, r and θ are variables and which is a constant. If in a certain case the radius r is constant, then the equation gives A in terms of θ and it is possible to find $dA/d\theta$. This is known as differentiating A with respect to θ. Alternatively, if the angle θ is constant, then we can differentiate A with respect to r to find dA/dr.

Standard forms for differentiation

The following standard forms may be needed for reference purposes in the solution of some problems: other standard forms are beyond the scope of this book.

$y = \text{constant}$	$\frac{dy}{dx} = 0$	$y = a^x$	$\frac{dy}{dx} = a^x \log a$
$y = x^n$	$\frac{dy}{dx} = nx^{n-1}$	$y = ax^n$	$\frac{dy}{dx} = anx^{n-1}$

Differentiation of a product

The rule is to differentiate one at a time, regarding the rest as constant, and then add the results together.

If $y = uv$ where u and v are both functions of x, then

$$\frac{dy}{dx} = u\frac{dv}{dx} + v\frac{du}{dx}$$

and the same procedure is applicable to products containing more terms.

Examples:

(a) $y = x(4 + x)^5$

$$\frac{dy}{dx} = x.5(4 + x)^4 + 1.(4 + x)^5$$

$$= 2(3x + 2)(4 + x)^4$$

(b) $y = \sqrt{x}(2 + x)^2$

i.e. $y = x^{1/2}(2 + x)^2$

$$\frac{dy}{dx} = x^{1/2}.2(2 + x) + \tfrac{1}{2}x^{-1/2}(2 + x)^2$$

$$= 2\sqrt{x}(2 + x) + \frac{(2 + x)^2}{2\sqrt{x}}$$

$$= \frac{4x(2 + x) + (2 + x)^2}{2\sqrt{x}}$$

$$= \frac{5x^2 + 12x + 4}{2\sqrt{x}}$$

(c) $y = \dfrac{x^2}{(2x + 5)^3}$

This is a quotient, but may be expressed in the form of a product, thus:

$$y = x^2(2x + 5)^{-3} \quad \frac{dy}{dx} = 2x(2x + 5)^{-3} + x^2(2x + 5)^{-4} \times -3 \times 2$$

$$= \frac{2x}{(2x + 5)^3} - \frac{6x^2}{(2x + 5)^4}$$

$$= \frac{2x(5 - x)}{(2x + 5)^4}$$

Differentiation of a quotient

If $y = u/v$, where u and v are both functions of x, then

$$\frac{dy}{dx} = \frac{v\frac{du}{dx} - u\frac{dv}{dx}}{v^2}$$

Examples:

(a) $y = \dfrac{2x}{1 + x}$ putting $u = 2x$, $\dfrac{du}{dx} = 2$

$v = 1 + x$, $\dfrac{dv}{dx} = 1$

Substituting in the formula

$$\frac{dy}{dx} = \frac{v\frac{du}{dx} - u\frac{dv}{dx}}{v^2}$$

gives the result

$$\frac{dy}{dx} = \frac{(1 + x).2 - 2x.1}{(1 + x)^2}$$

which simplifies to

$$\frac{dy}{dx} = \frac{1}{(1 + x)^2}$$

(b) $y = \dfrac{x^2}{(2x + 5)^3}$ putting $u = x^2$, $\dfrac{du}{dx} = 2x$

$v = (2x + 5)^3$, $\dfrac{dv}{dx} = 6(2x + 5)^2$

Substituting,

$$\frac{dy}{dx} = \frac{(2x + 5)^3 . 2x - 6x^2(2x + 5)^2}{(2x + 5)^6}$$

$$= \frac{(2x + 5) . 2x - 6x^2}{(2x + 5)^4} = \frac{10x - 2x^2}{(2x + 5)^4}$$

$$= \frac{2x(5 - x)}{(2x + 5)^4}$$

(yielding the same result as in Example (c) above).

Successive differentiation

If y is a function of x, we can differentiate to find dy/dx which is the differential coefficient of y with respect to x. If we differentiate again we shall have the second differential coefficient of y with respect to x twice and, in the shorthand used in mathematics, this is denoted by the symbol d^2y/dx^2. Similarly, we could continue the process and obtain d^3y/dx^3, $d^4y/dx^4, \ldots$ d^ny/dx^n.

Examples: If $y = x^n$, $\dfrac{dy}{dx} = nx^{n-1}$, $\dfrac{d^2y}{dx^2} = n(n - 1)x^{n-2}$, $\ldots \dfrac{d^ny}{dx^n} = \underline{|n}$

If $y = e^{2x}$, $\dfrac{dy}{dx} = 2e^{2x}$, $\dfrac{d^2x}{dx^2} = 4e^{2x}$, $\ldots \dfrac{d^ny}{dx^n} = 2^ne^{2x}$

To save space, mathematicians sometimes abbreviate even further and denote successive differential coefficients of y with respect to x as $y', y'', \ldots y^n$, but it is not advised that you should adopt this practise.

Other equations can also be differentiated more than once, and a distance/time relationship giving s in terms of t will yield first ds/dt and then d^2s/dt^2. Even this can be further abbreviated and first and second differential coefficients with respect to t are denoted by dots placed over the symbol of the dependent variable, for example

$$\dot{s} = \frac{ds}{dt}, \quad \ddot{s} = \frac{d^2s}{dt^2}, \quad \dot{\theta} = \frac{d\theta}{dt}$$

EXERCISE 9(b)

1. If $y = 3x^2 + 4$, find dy/dx.
2. If $r = l/\theta$ where l is a constant, find $dr/d\theta$.
3. If $pv^n = c$, where c and n are constants, prove that $v^{n-1}\dfrac{dp}{dv} + cn = 0$.

4. if $s = 245t - 4{\cdot}9t^2$, find the values of $\frac{ds}{dt}$ and $\frac{d^2s}{dt^2}$.

   Deduce the value of t which makes $\frac{ds}{dt} = 0$.

5. If $y = x^3 - \frac{1}{1-x}$, find $\frac{d^2y}{dx^2}$.

6. Differentiate the following:

   (a) $y = 9x(2 - x)^4$ (b) $y = x^2\sqrt{(1 + 2x)}$ (c) $y = x^3\sqrt{(4 - 3x)}$

   (d) $y = \frac{3x - 2}{(5 - 4x)^2}$ (e) $y = \frac{x^2}{\sqrt{(4 - x)}}$

Integration

Starting with an equation expressing y in terms of x, we have seen how to differentiate y with respect to x to obtain the differential coefficient denoted by the symbol dy/dx. We shall now consider the reverse process. The symbol

$$\int y\,dx$$

may be interpreted as 'the integral of y with respect to x' and its value will be an expression which becomes y when differentiated.

Examples:

(1) Differentiating x^3 yields $3x^2$, so

$$\int 3x^2\,dx = x^3$$

(2) Differentiating $(1 + 3x)^4$ yields $12(1 + 3x)^3$, so

$$\int 12(1 + 3x)^3\,dx = (1 + 3x)^4$$

and it follows that

$$\int (1 + 3x)^3\,dx = \frac{1}{12}(1 + 3x)^4$$

(3) Differentiating ax yields a, so

$$\int a\,dx = ax$$

whatever the value of the constant a.

Arbitrary constant

Since any constant becomes zero upon differentiation, in the reverse process of integration we must allow for the possible presence of a constant.

In some cases the precise value of the constant can be determined if certain values are known, but in every other case an unknown constant must be added. This is usually denoted by the letter C and is often referred to as the **arbitrary constant.**

In the above examples the arbitrary constant has been omitted and it would have been better to have written the results as follows:

$$\int 3x^2\,dx = x^3 + C$$

$$\int (1 + 3x)^3\,dx = \frac{1}{12}(1 + 3x)^4 + C$$

$$\int a\,dx = ax + C$$

General form

We can obtain x^n by differentiating

$$\frac{x^{n-1}}{n + 1} + C$$

and thus, in reverse,

$$\int x^n\,dx = \frac{x^{n+1}}{n + 1} + C$$

This result is true for all values of n, including negative and fractional values, and its use is illustrated in the following examples:

(1) $\int x^4\,dx = \dfrac{x^5}{5} + C$

(2) $\int \dfrac{dx}{x^3} = \int x^{-3}\,dx = \dfrac{x^{-2}}{-2} + C = -\dfrac{1}{2x^2} + C$

(3) $\int \sqrt{x}\,dx = \int x^{\frac{1}{2}}\,dx = \dfrac{x^{\frac{3}{2}}}{3/2} + C = \frac{2}{3}\sqrt{x^3} + C$

It should be noted that when a power of x is multiplied by a constant, that constant remains unchanged by the process of integration, and can be placed outside the integration sign.

e.g. $$\int 6x^2\,dx = 6\int x^2\,dx = \frac{6 \times x^3}{3} + C = 2x^3 + C$$

Exception to the general form

The one exception to the general form above is the case where $n = -1$. In this case the integral is a logarithmic form and is unlikely to occur at this stage.

Examples:

(1) $$\int \frac{dx}{x} = \log_e x + C$$

(2) $$\int \frac{dx}{1 + x} = \log_e(1 + x) + C$$

(3) $$\int \frac{dx}{a + bx} = \frac{1}{b} \log_e(a + bx) + C$$

Standard integrals

Other standard forms involving trigonometrical or exponential functions are also outside the scope of this book but will be dealt with in a subsequent volume.

Integration of a sum

When a sum of several terms is integrated the result will be the sum of their separate integrals. Only one arbitrary constant is required.

Examples:

(1) $$\int (3x^2 - 8x + 4)\, dx = x^3 - 4x^2 + 4x + C$$

(2) $$\int (5x - \sqrt{x})^2\, dx = \int (25x^2 + 10x^{\frac{3}{2}} + x)\, dx$$
$$= \frac{25x^3}{3} + 4x^{\frac{5}{2}} + \frac{x^2}{2} + C$$

EXERCISE 9(c)

1. Integrate with respect to x:
   (a) $4x$ (b) $8x^3$ (c) $\sqrt{x}$ (d) $x^{0.15}$

2. If $dy/dx = 9x^2 + 6x - 10$, find y, given that $y = 0$ when $x = 1$.

3. Given that $s = 0$ when $t = 0$, and $\dfrac{ds}{dt} = 49 - 9.8t$, find s in terms of t. Hence determine the value of s when $\dfrac{ds}{dt} = 0$.

4. Integrate:

(a) $\int \frac{dx}{2x^3}$ (b) $\int \frac{dx}{x^{2.5}}$ (c) $\int \frac{4}{\sqrt{x}}\, dx$ (d) $\int \frac{(x+3)}{x^4}\, dx$

5. Integrate:

(a) $\int (3-2x)^3\, dx$ (b) $\int \sqrt{1+4x}\, dx$

6. Integrate:

(a) $\int 4\pi r^2\, dr$ (b) $\int 720v^{1.4}\, dv$ (c) $\int (4-0.8t^3)\, dt$

7. Evaluate the integrals:

(a) $\int_0^3 (1+3x)^3\, dx$ (b) $\int_0^3 \frac{dx}{(3+2x)^3}$

REVISION EXERCISE 9

1. (a) If $s = 16t^2$ find values of s when $t = 2.$ 2.1, 2.01, 2.001, and $2 + \delta t$. Hence find the changes δs in s corresponding to changes δt in t of 0.1, 0.01, 0.001 and δt, when $t = 2$. Calculate $\delta s/\delta t$ for these four pairs of results. Calculate ds/dt from the last pair.
   (b) Differentiate: (i) $3\sqrt{t^3}$; (ii) $(3x+2)^3$.
   (c) Calculate the gradient of the curve $y = -x^3 + 4x^2 - 3x$ at each of the points where it crosses the x-axis. (E.M.E.U.)

2. (a) Differentiate from first principles $12x^2 + 2$.
   (b) Differentiate:

   (i) $2x^5 - \frac{1}{2}\sqrt{x} + \frac{1}{\sqrt{x}}$;

   (ii) $\frac{x^2+3}{x-4}$;

   (iii) $(x^2+3)\sqrt{x+1}$. (N.C.T.E.C.)

3. (a) Differentiate:

   (i) $4x^{-1}$ (ii) $3x^6$ (iii) $8x^{10}$
   (iv) x^3 (v) $(7x - 8x^2)$

   (b) Integrate the following functions:

   (i) $6x + 2$ (ii) $9x^2 + 8x^5$ (iii) $7x$
   (iv) $9(x+2)$ (v) x^{10} (N.C.T.E.C.)

4. Differentiate the following functions and simplify as far as possible:

   (i) $\sqrt{(1-x)}$ (ii) $\frac{1}{2x^3}$ (iii) $\frac{4x}{1-3x}$ (U.E.I.)

5. Find the average value of the gradient of the curve $y = \frac{2}{3}x^2 + 8$ between the pair of points (a) where $x = 2$ and $x = 2.1$, and (b) where $x = 2$ and $x = 2.01$. If this process was repeated with the second value each time getting closer and closer to 2, what would the gradient finally become? (U.L.C.I.)

6. What is the average value of the gradient of the curve $y = \frac{1}{2}x^2 - 2$ between each of the pairs of points (a) $x = 3$ and $x = 3.1$ and (b) $x = 3$ and $x = 3.01$? From these results give an estimate of the value of the gradient at the point $x = 3$. (U.L.C.I.)

7. (a) Plot the graph of $y = x^3$ for values of x from 0 to 4 and, by direct measurement, determine
   (i) the average value of $\delta x/\delta x$ when x increases from 2 to 3 and δy, δx, are the respective increments in y, x.
   (ii) the value of dy/dx, the gradient, when $x = 2.5$.

   (b) Differentiate the following functions with respect to x:
   (i) $y = 3x^4 - 1/x^5$ (ii) $y = (1-x)^3/x$ (iii) $y = 8x^{-2\cdot3} \times \sqrt{3x^{4\cdot4}}$
   (N.C.T.E.C.)

8. (a) Integrate the following functions with respect to x.
   (i) $y = x^{-7} - \dfrac{10}{\sqrt{x}} + 5$ (ii) $y = \sqrt{3x^{-0\cdot9}}$ (iii) $y = 3x^2 + \dfrac{1}{4}x^{1/4}$

   (b) Find y in terms of x if $\dfrac{dy}{dx} = x^5 - x^2$ and $y = 0$ when $x = 1$; find also the values of dy/dx when $y = \frac{1}{6}$. (N.C.T.E.C.)

10 Applications of Differentiation

Velocity and acceleration

The equation $s = 7t - 4{\cdot}9t^2$ gives the distance s metres travelled in t seconds by a stone thrown upwards with an initial velocity of 7 m/s. Differentiating this equation gives an expression for ds/dt which is the gradient at any point on the distance-time graph. Since this is a measure of the change of distance with time, it does in fact represent the velocity.

$$s = 7t - 4{\cdot}9t^2$$

$$v = \frac{ds}{dt} = 7 - 9{\cdot}8t$$

This derived equation, $v = 7 - 9{\cdot}8t$, gives us the velocity of the stone at any time. After $\frac{1}{2}$ second, $v = 7 - 4{\cdot}9 = 2{\cdot}1$ m/s; after 1 second, $v = 7 - 9{\cdot}8 = -2{\cdot}8$ m/s, which indicates that the stone is then descending. To find out just when the stone reaches the highest point of its flight, we put $v = 0$ since the stone at that instant has no velocity.

Then
$$0 = 7 - 9{\cdot}8t$$

$$t = \tfrac{5}{7} \text{ seconds}$$

To find the highest point reached by the stone, we put this value into the original expression

$$s = 7t - 4{\cdot}9t^2$$

$$= 5 - 2{\cdot}5 = 2{\cdot}5 \text{ m}$$

If we take the equation

$$v = \frac{ds}{dt} = 7 - 9{\cdot}8t$$

and differentiate again, we shall get the rate of change of velocity, i.e. the acceleration

$$\frac{dv}{dt} = \frac{d^2s}{dt^2} = -9{\cdot}8$$

which reveals that the stone is subject to a retardation of $9{\cdot}8$ m/s^2.

Equations of motion

In the example above, the acceleration was due to gravity and this acceleration is constant both in its value of 9·8 m/s^2 and in its downward direction. For any motion with constant acceleration (or retardation) the following equations apply.

$$s = ut + \tfrac{1}{2}at^2$$

$$v = u + at$$

$$v^2 = u^2 + 2as$$

where s is the distance travelled (in metres)
t is the time (in seconds)
u is the initial velocity (in m/s)
v is the velocity after time t
a is the constant acceleration (in m/s^2)

Example: A vehicle starting from rest covers 62·5 m in the first 5 seconds. Assuming the acceleration to be constant, find the speed at this time and the distance which should be covered in the next five seconds.

Using $s = ut + \frac{1}{2}at^2$, and substituting $s = 62{\cdot}5$, $t = 5$, $u = 0$,

$$62{\cdot}5 = 0 + \tfrac{1}{2}a.5^2$$

$$a = 5\ \text{m/s}^2$$

Using $v^2 = u^2 + 2as$, $v^2 = 0 + 625$, $v = 25$ m/s
i.e. after the first seconds the speed is 25 m/s.
Using $s = ut + \frac{1}{2}at^2$ again for the second part, $t = 5$, $u = 25$, $a = 5$.

$$s = 125 + 62{\cdot}5 = 187{\cdot}5\ \text{m}$$

i.e. in the next five seconds, the vehicle should cover a further 187·5 m.

Note how the final velocity of the first stage becomes the initial velocity of the second stage. Alternatively, the problem could be tackled from the start each time and the distance covered in the second interval of time would then be given by the difference between the distance covered in the first five seconds and the distance covered in the first ten seconds.

Theoretically, the equations of motion given above apply to particles, but they are equally valid for larger bodies, even large objects such as trucks, provided that the motion being considered is in a straight line and no rotation of the body need be taken into account. The motion of bodies which are rotating, sliding, or toppling over, is outside the scope of this book.

Graphical relationships—maxima and minima

The equation $x^3 - 4x^2 - 11x + 30 = 0$ is a cubic and will have three solutions. The graph of $y = x^3 - 4x^2 - 11x + 30$ is shown in fig. 46(a) and crosses the x-axis at three points corresponding to the three solutions of the cubic, $x = -3, 2, 5$. As the curve is continuous, it must change direction in order to recross the x-axis, and, as can be seen from the graph, it rises to a maximum where $x = -1$ and falls to a minimum at $x = 3\frac{2}{3}$. It is also apparent that the curvature changes its direction at $x = 1\frac{1}{3}$. Differentiating the equation:

$$y = x^3 - 4x^2 - 11x + 30$$

we get
$$\frac{dy}{dx} = 3x^2 - 8x - 11$$

and
$$\frac{d^2y}{dx^2} = 6x - 8$$

The graphs of $y = 3x^2 - 8x - 11$ and $y = 6x - 8$ are shown in figs. 46(b) and 46(c).

From fig. 46(b), it can be seen that the solutions of the equation

$$3x^2 - 8x - 11 = 0$$

are the turning points of the curve $y = x^3 - 4x^2 - 11x + 30$, and this indicates the general rule that turning points may be found by differentiating the equation of the curve and equating dy/dx to zero. Furthermore, it is clear from the graph that maxima and minima may be distinguished by reference to the sign of dy/dx in the neighbourhood of the x-values thus obtained. In the case under consideration, if

$$\frac{dy}{dx} = 0$$

then
$$3x^2 - 8x - 11 = 0$$

i.e.
$$(x + 1)(3x - 11) = 0$$

and thus
$$x = -1 \text{ or } 3\tfrac{2}{3}$$

Around the point where $x = -1$ in the fig. 46(b), the value of the expression $3x^2 - 8x - 11$ changes sign from positive to negative, i.e. the value of dy/dx for the expression $y = x^3 - 4x^2 - 11x + 30$ changes sign from positive to negative, and, as can be seen on the graph above, this indicates a maximum value of the expression $x^3 - 4x^2 - 11x + 30$. Similarly, when $x = 3\frac{2}{3}$, the sign of dy/dx changes from negative to positive, and this indicates a minimum value of $x^3 - 4x^2 - 11x + 30$.

From fig. 46(c), we see that the x-axis is crossed at a point which corresponds to a turning point on fig. 46(b) and a point of inflexion on fig. 46(a). The point of inflexion is where the curvature changes direction, and it is now apparent that it may be found by differentiating the equation of the curve twice and equating $d\,y/dx^2$ to zero. In addition, we see from fig. 46(c) that, when the original expression has a maximum value, the sign of d^2y/dx^2 is negative, and, conversely, for a minimum it is positive: this provides us with a useful alternative method of distinguishing between maxima and minima.

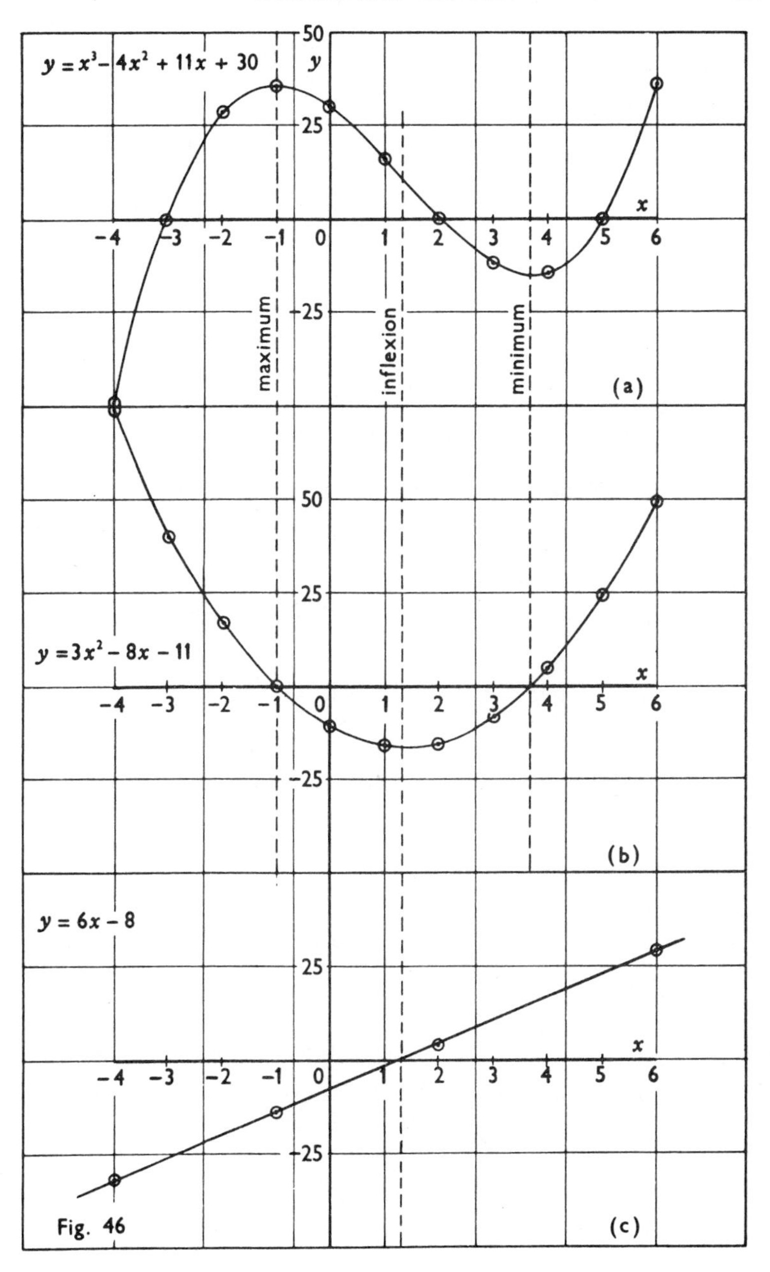

$y = x^3 - 4x^2 + 11x + 30$
y
50
25
-25
x
-4
-3
-2
-1
0
1
2
3
4
5
6
maximum
inflexion
minimum
(a)
50
25
$y = 3x^2 - 8x - 11$
-25
(b)
$y = 6x - 8$
25
-25
(c)

Fig. 46

Example: If

$$y = x^3 + x^2 - 8x - 3$$

$$\frac{dy}{dx} = 3x^2 + 2x - 8 = (3x - 4)(x + 2)$$

For turning points,

$$\frac{dy}{dx} = 0$$

$$\therefore \quad x = +1\tfrac{1}{3} \text{ or } -2$$

$$\frac{d^2y}{dx^2} = 6x + 2$$

which is positive when $x = 1\frac{1}{3}$, but negative when $x = -2$.
When $x = 1\frac{1}{3}$, $y = -6\frac{22}{27}$; when $x = -2$, $y = 9$. Our final conclusion is therefore that the curve has a minimum at the point $(1\frac{1}{3}, -6\frac{22}{27})$ and a maximum at the point $(-2, 9)$.

Problems involving maxima and minima

The principles outlined above may be used in the solution of many practical problems.

Example 1 An isolated rectangular building site of 800 m² is to be fenced off from the rest of a large meadow. If the cost per metre of fencing for the frontage is three times as much as the cost per metre of fencing for the remainder, calculate the dimensions for which the cost of fencing is a minimum.

Let the frontage of the plot be f metres and the depth d metres.
Then the area of the plot:

$$fd = 800$$

$$\therefore \quad d = \frac{800}{f}$$

The cost of fencing:

$$C = 3f + d + f + d = 4f + 2d$$

Now we cannot differentiate whilst C is represented in terms of the *two* variables, f and d, but must substitute to eliminate one of them.

$$C = 4f + 2\left(\frac{800}{f}\right)$$

$$\frac{dC}{df} = 4 - \frac{1600}{f^2}$$

For a turning point $$\frac{dC}{df} = 0$$

$$\therefore \quad 4 - \frac{1600}{f^2} = 0$$

$$f^2 = 400$$

$$f = 20 \text{ m and } d = 40 \text{ m}$$

i.e. for the cost of fencing to be a minimum, the site should have a frontage of 20 m and a depth of 40 m.

Example 2 Find the ratio of height to radius for cylindrical cans to be manufactured from the minimum quantity of thin sheet metal.

Let the height be h mm, the radius be r mm, the volume V mm³.

Note that for any particular capacity of can, the volume V is constant.

For the volume, $V \pi r^2 h$ and hence $h = V/\pi r^2$

For the area of metal, $A = 2\pi r^2 + 2\pi rh$

Substituting for h, $$A = 2\pi r^2 + \frac{2V}{r}$$

Differentiating, $$\frac{dA}{dr} = 4\pi r + \frac{2V}{r^2}$$

For a minimum value, $dA/dr = 0$,

i.e. $$4\pi r - \frac{2V}{r^2} = 0$$

Substituting for V,

$$4\pi r - \frac{2\pi r^2 h}{r^2} = 0$$

$$4\pi r - 2\pi h = 0$$

$$h = 2r$$

i.e. for minimum area of sheet metal, the ratio of height to radius should be 2 to 1.

EXERCISE 10

1. Assuming the acceleration due to gravity to be 9·8 m/s (downwards), find the greatest height attained by a bullet fired straight up into the air with a velocity of 42 m/s.

2. Find the velocity acquired by a brick carelessly allowed to fall from a height of 10 m.

3. An object dislodged from some scaffolding seriously injured a man below. From the mass of the object and the depth of penetration, it was deduced that it was travelling at 35 m/s when it struck the man below. Find how many metres it fell.

4. A vehicle travelling at 30 m/s can be brought to a halt in 60 m. Assuming the retardation produced by the brakes is uniform, find the braking distance required when the same vehicle is travelling at 36 m/s.

5. If the equation $\theta = 16\pi + 28t - 2t^2$ gives the angle in radians through which a wheel turns in t seconds, find after how many seconds the wheel comes to rest. Calculate the angle turned through in the last second of movement.

6. Find the maximum value of $x(6-x)$.

7. Find the points at which the curve $y = x^3 - 3x^2 - 24x + 20$ has its maximum and minimum values.

8. Locate the turning points on the following curves and distinguish between maxima and minima.

   (a) $y = x + \frac{4}{x}$ (b) $y = 16x + \frac{1}{x^2}$ (c) $y = 1 + \frac{1}{x} - \frac{1}{x^2}$

9. Find where the curve $y = 12 + 8x - 5x^2 + x^3$ cuts the x-axis and locate the turning points.

10. Find the nature and position of the turning points on the following curves:

    (a) $y = 9 + 4x - x^2$ (b) $y = 6x^4 - 16x^3 - 21x^2 + 30x - 5$.

11. A square sheet of metal of side 300 mm has squares of side x mm removed from the corners, so that an open box may be formed. Find the value of x such that the volume of the box is a maximum.

12. A tunnel is to be constructed with a flat floor and vertical side walls surmounted by a semi-circular curved roof. If d is the diameter of the semi-circle and h is the height of the side wall, find the ratio of d to h for the perimeter of the cross-section to be a minimum.

13. An open water trough is to be constructed from three long boards of equal width. One board is laid flat for the base and the others form sloping sides each at an angle of x radians to the vertical. Find the value of x for the trough to have maximum capacity.

14. A log of circular cross-section tapers uniformly from 45 mm diameter at one end to 25 mm diameter at the other end. If the length of the log is 4 m, find the length of the largest beam with square cross-section which may be cut from this log.

REVISION EXERCISE 10

1. (a) A square plate of sheet metal has a side of length 15 units. A rectangular tray without a lid is made from this sheet by cutting squares of side x units from the corners and folding and welding the vertical edges. Show that the volume V units^3 of such a tray is given by

$$V = 225x - 60x^2 + 4x^3 .$$

Find the maximum volume of the tray, proving it is a maximum.

(b) A body at time t seconds is distant x m from a fixed point 0 given by $x = 10(1 - \cos 2t)$. Find the first two times when (i) the body is at rest, (ii) the acceleration is zero. Calculate the velocities of the body when the acceleration is zero. (E.M.E.U.)

2. (a) Differentiate the following with respect to x:

(i) $(x^2 - 3x + 5)(x^2 + 2x - 3)$

(ii) $\dfrac{x-7}{2x+3}$

(iii) $(2x^3 - 3)^6$

(b) The equation of a curve is $y = 2x^3 - 9x^2 + 12x + 6$. Use calculus to find its turning points, showing clearly the nature of each point and why.
Sketch the curve. What is the gradient of the curve where it crosses the y axis? (U.E.I.)

3. (a) Differentiate the following functions with respect to x, simplifying first where necessary.

(i) $1/(4\pi x)^3$ (ii) $-0{\cdot}32x^{\frac{1}{2}} - 2{\cdot}4x^{-\frac{1}{2}} - 9{\cdot}6$ (ii) $(3x^2 - \frac{1}{3})(\frac{1}{3}x^{-2} + 3)$.

(b) The height, h, in metres, of an object t seconds after being fired into the air is given by $h = 19{\cdot}2t - 4{\cdot}8t^2$.
Find (i) the initial velocity of the object (i.e. at $t = 0$)
and (ii) the height it has attained when its velocity is one half of its initial velocity. (N.C.T.E.C.)

4. (a) Differentiate the following expressions and simplify as far as possible:

(i) $\sqrt{(1 + x^2)}$ (ii) $\dfrac{3}{x^4}$ (iii) $\dfrac{x^2}{1 - x}$

(b) Calculate the coordinates of the maximum and minimum points on the curve $y = 2x^3 - 6x^2 + 1$ and distinguish between them. Plot the graph of the curve between the values of $x = -1$ and $x = +3$. (U.E.I.)

5. (a) The ordinate at the point of intersection of the curves $y = 4 - x^3$ and $y = \frac{1}{3}(x+2)^2$ is 3. Calculate the slope of each curve at this point of intersection.

(b) Find the co-ordinates of the turning points of the curve

$$y = \frac{x^3}{3} + \frac{x^2}{2} - 6x + 9 .$$

Hence find the maximum and minimum values of $\frac{1}{3}x^3 + \frac{1}{2}x^2 - 6x + 9$.

(W.J.E.C.)

6. A 10 m long horizontal beam AB is freely supported at its ends. It has a varying cross-section so that its deflection y m at a point x m along the beam from A is given by the formula $y = \frac{1}{3000}(x^3 - 26x^2 + 160x)$.
Determine the point on the beam at which the deflection is a maximum and calculate this deflection in millimetres. (W.J.E.C.)

7. (a) Explain how to find the slope of the tangent at any point on the curve $y = x + \frac{1}{x}$. For what values of x is the slope zero? Illustrate by means of a rough sketch of the curve.
(b) Differentiate the following:
(i) $(2x + 3)^5$ (ii) $\frac{3}{\sqrt[3]{x^2}}$
(c) A sheet of metal measures 240 mm by 90 mm. Four equal squares are cut from the sheet, one from each corner. The sides are then turned up to make an open box. Find the size of the squares to make the box with the greatest volume. (E.M.E.U.)

8. (a) Differentiate with respect to x and simplify:
(i) $\frac{1}{3x^2}$ (ii) $\sqrt{1 + x^2}$ (iii) $\frac{3x}{1 - 2x}$
(b) An open gutter made from strip metal, 40 mm wide, has a rectangular cross-section. Find the width and depth which give the maximum area of cross-section.

9. (a) Differentiate:
(i) $\frac{x^2}{1 - x}$ (ii) $\frac{1}{2x^5}$ (iii) $(1 - 3x)^{\frac{1}{3}}$
(b) Given the equation $y = \frac{2}{3}x^3 + x^2 - 12x + 3$, find, using the calculus, the maximum and minimum values of y. (W.J.E.C.)

10. (a) Differentiate with respect to x the expressions:
(i) $y = (2x + 3)(4 - 3x^2)$ (ii) $\frac{3x^2}{2x - 5}$
(b) A rectangular tank has a square base and an open top and a capacity of 4 m^3. Show that if a side of the base is x m then the total area of sheet metal used (ignoring laps) in the construction of the tank is given by:
$$A = x^2 + \frac{16}{x}$$
Then determine:
(i) the value of x for which A is a minimum;
(ii) the minimum value of A. (N.C.T.E.C.)

11. (a) Obtain expressions for dy/dx for:
(i) $y = 3x(x + 2)^7$ (ii) $y = \frac{2x - 1}{x + 1}$
(b) Assuming a paint drum to be a simple closed cylinder, find the value of
$$\frac{\text{height of drum}}{\text{diameter of drum}}$$
which will give the greatest volume for a fixed amount of sheet metal.
(N.C.T.E.C.)

12. The deflection y at a point distant x from the end of a beam is given by the expression:

$$y = \frac{1}{EI}\left[\frac{3x^3}{4} - \frac{x^4}{50} - \frac{400x}{3}\right]$$

The bending moment at the point x is given by:

$$M = EI\frac{d^2y}{dx^2}$$

Find:

(a) the value of M when $x = 4$;

(b) the value of x when M is a maximum. (K.C.E.A.B.)

13. The deflection, y, at a distance, x, from the fixed end of a cantilever is given by the equation

$$y = \frac{W}{EI}\left(\frac{lx^2}{2} - \frac{x^3}{6}\right)$$

where W, E, I and l are constants. Obtain d^2y/dx^2, and find its value when $x = \frac{1}{3}l$. Hence find the corresponding value of the bending moment M, given that

$$M = EI\frac{d^2y}{dx^2}$$

(W.J.E.C.)

14. (a) Find the values of x at the turning points on the graph of

$$y = 4x^3 + 9x^2 - 12x + 3$$

and determine whether these are maximum or minimum points on the curve.

(b) The strength of a beam of rectangular section is directly proportional to the breadth and the square of the depth. Find the dimensions of the strongest beam section that can be cut from a circular log of 45 mm diameter. (U.L.C.I.)

11 Applications of Integration

Definite integrals—area under a curve

In fig. 47 we wish to find the area under the curve between the limits $x = a$ and $x = b$. Consider a strip of width δx (where δx is very small) at a distance x from the y-axis. The area of this narrow strip is approximately $y \,.\, \delta x$ and the total area of the section required is given approximately

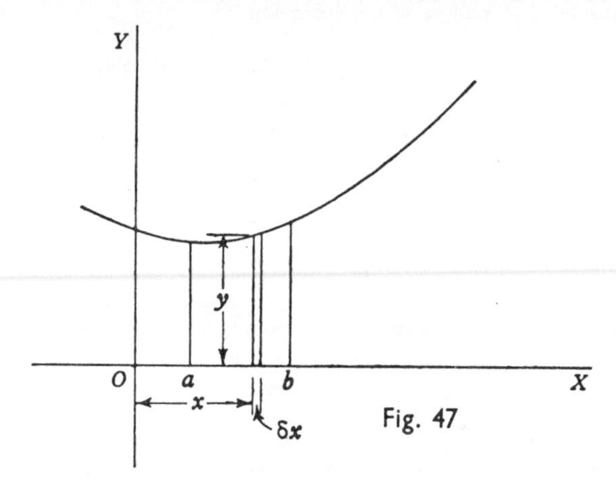

Fig. 47

by the sum of all such strips, of area $y \,.\, \delta x$, from $x = a$ to $x = b$. Now the accuracy would be increased by increasing the number of strips, i.e. by reducing the value of δx. Taking this to the limit, the area is given accurately by an infinite number of strips of which the width δx has become the infinitesimal dx. For the sum of these strips we use the long s which has already been introduced as the symbol of integration and the limits are placed at the head and foot of this symbol. In this notation,

$$\text{area} = \int_a^b y \, dx$$

b and a are referred to as the upper and lower limits and, in practice, after integration these values are substituted in the expression in turn, the difference being the area.

Example:

Find the area between the curve $y = x^2$ and the x-axis between the limits $x = 1$ and $x = 4$.

$$\text{area} = \int_1^4 x^2\,dx$$

$$= \left[\frac{x^3}{3}\right]_1^4 = \frac{64}{3} - \frac{1}{3} = 31 \text{ units}^2$$

Areas above the axis come out positive, whilst the part below the axis comes out negative. This illustrates the necessity of checking whether or not a given curve crosses the axis within the given limits.

Note that with a definite integral no arbitrary constant is required: if inserted it would in any case cancel out.

Approximate methods for the evaluation of integrals

In chapter 7, various methods were proposed for finding the areas of irregular figures. Each of these methods can be used to estimate the area under a particular curve. For Simpson's Rule and for the Trapezoidal Rule it is not even necessary to draw the corresponding curve. The interval between the limits of integration should be divided into an appropriate number of equally spaced smaller intervals and the corresponding ordinates calculated. The method is illustrated by the example on page 95.

Volume of a solid of revolution

If the area between a given curve and the x-axis be rotated about the x-axis for a complete revolution, the volume swept out will be given by

$$\pi \int y^2\,dx$$

between any required limits.

By selecting a suitable straight line for the given curve, it is possible to determine the volume of a cylinder, a cone, or the frustum of a cone. If the given curve is a circular arc, e.g. part of the curve $y^2 = a^2 - x^2$, then it can be used to find the volume of a cap or a zone of a sphere.

Example: The tapering end of a tool is in the form of a solid of revolution obtained by the rotation about the x-axis of the area under the curve $y = \frac{1}{8}x^{\frac{3}{2}}$ from $x = 0$ to $x = 4$. Find its volume. (W.J.E.C.)

$$V = \pi \int y^2\,dx = \frac{\pi}{64}\int_0^4 x^3\,dx = \frac{\pi}{64}\left[\frac{x^4}{4}\right]_0^4$$

$$= \frac{\pi}{64}\left[\frac{4^4}{4} - 0\right] = \pi \text{ cu. units.}$$

Application of integration to the deflection of beams

The bending moment M at any point on a beam under given conditions may be found by taking moments in the usual way, and since

$$\frac{d^2y}{dx^2} = \frac{M}{EI}$$

and both E and I are constants, the deflection y, in terms of x, may be found by integrating twice.

Example 1 Find an expression for the deflection at a distance x from the free end of a cantilever of length l fixed at one end and carrying a load W at its free end.

In this case, the bending moment is proportional to the distance along the beam, and

$$M = -Wx$$

$$EI\frac{d^2y}{dx^2} = -Wx$$

Integrating, $$EI\frac{dy}{dx} = A - \frac{Wx^2}{2}$$

Now $\frac{dy}{dx} = 0$ when $x = l$, since this end is fixed level, $\therefore \quad A = \frac{Wl^2}{2}$

$$EI\frac{dy}{dx} = \frac{W}{2}(l^2 - x^2)$$

Integrating, $$EIy = \frac{W}{2}\left(l^2x - \frac{x^3}{3}\right) + B$$

Now $y = 0$ when $x = l$, since there is no deflection at the fixed end,

$$\therefore \quad B = -\frac{Wl^3}{3}$$

$$EIy = \frac{W}{6}(3l^2x - x^3) - \frac{Wl^3}{3}$$

$$y = \frac{W}{6EI}(3l^2x - x^3 - 2l^3)$$

Example 2 Find an expression for the deflection at any point on a simply supported beam of length l carrying a uniformly distributed load w per unit length.

In this case it is possible to start from the relation between the loading w and the shearing force F: $\frac{dF}{dx} = -w$.

The relation between shearing force and bending moment is $\frac{dM}{dx} = F$.

From these we derive that $\frac{d^2M}{dx^2} = -w$

Integrating, $$F = -wx + A$$

Now $F = 0$ when $x = l/2$ (by symmetry) $\therefore \quad A = wl/2$

$$\frac{dM}{dx} = \frac{wl}{2} - wx$$

Integrating, $$M = \frac{wlx}{2} - \frac{wx^2}{2}$$

and in this case the arbitrary constant is zero, since $M = 0$ when $x = 0$. (The same expression for M may be derived by taking moments if the load over the section of length x is assumed to act at its centre of gravity)

$$EI\frac{d^2y}{dx^2} = \frac{wlx}{2} - \frac{wx^2}{2}$$

Integrating, $$EI\frac{dy}{dx} - \frac{wlx^2}{4} - \frac{wx^3}{6} + B$$

But the beam must be level at the centre, $\therefore$ when $x = \frac{l}{2}, \frac{dy}{dx} = 0$

$$\therefore \quad B = -\frac{wl^3}{24}$$

$$EI\frac{dy}{dx} = \frac{wlx^2}{4} - \frac{wx^3}{6} - \frac{wl^3}{24}$$

Integrating, $$EIy = \frac{wlx^3}{12} - \frac{wx^4}{24} - \frac{wl^3x}{24}$$

and in this case the arbitrary constant is again zero, since there can be no deflection at the support, so $y = 0$ when $x = 0$.

$$\therefore \quad y = \frac{w}{24EI}(2lx^3 - x^4 - l^3x)$$

Note that the deflection is a maximum at the centre and its value there is

$$\frac{5wl^4}{384EI}$$

EXERCISE 11

1. Evaluate the following integrals:

(i) $\int_0^5 x^4 dx$ (ii) $\int_2^3 (3x^2-1)\,dx$

2. Evaluate:

(a) $\int_1^2 (x-1)(2-x)\,dx$ (b) $\int_2^3 \left(x+\frac{1}{x^2}\right) dx$ (E.M.E.U.)

3. (a) Evaluate:

(i) $\int_0^2 (x^3 - x^2 + x)\,dx$ (ii) $\int_2^3 \frac{dx}{x^2}$

(b) Find the values of:

(i) $\int (1 - x^2)^2\,dx$ (ii) $\int x(x^2 - 1)\,dx$ (N.C.T.E.C.)

4. Find the area between the curve $y = x^3$, the ordinates at $x = 1{\cdot}5$ and $x = 4$ and the x-axis. (K.C.E.A.B.)

5. On squared paper draw the curve $y = 6x - x^2$. Find the area bounded by the curve and the x-axis:
   (a) by integration;
   (b) by Simpson's Rule. (K.C.E.A.B.)

6. Find the area enclosed between $x = 0$, $x = 6$, the x-axis, and the line $y = x/3$. Find also the volume given by rotating this area about the x-axis.

7. The bending moment M at a distance x from one end of a beam is given by $M = 12x-x^2$. Find an expression for the deflection, y, in terms of x, given that $\frac{d^2y}{dx^2} = \frac{M}{EI}$, where E and I are constants, and that $\frac{dy}{dx} = 0$ when $x = 6$, and $y = 0$ when $x = 0$. (W.J.E.C.)

8. A uniform cantilever AB of length L and uniform weight w per unit length is clamped at the end A. The deflection of the cantilever from the horizontal, y, at a point distance x along it from A is given by the equation

$$\frac{d^2y}{dx^2} = \frac{w}{2EI}(L-x)^2$$

Find an expression, in ascending powers of x, for the deflection of the cantilever at the general point, $(x, y,)$ and determine the deflection and slope at the free end, B. (N.C.T.E.C.)

REVISION EXERCISE 11

1. (a) Integrate with respect to x:

$$\sqrt{x^3}+\frac{2}{\sqrt{x}}+\sqrt{x}$$

(b) Evaluate the definite integral:

$$\int_a^{2a} \left(\frac{a^3}{x^2}+\frac{x^2}{a}\right) dx$$ (U.L.C.I.)

2. Evaluate the following integrals in each case correct to three significant figures:

(i) $\displaystyle\int_1^2 \frac{dv}{v^{1\cdot 4}}$ (ii) $\displaystyle\int_0^2 e^{-1\cdot 2t}\, dt$ (iii) $\displaystyle\int_1^4 \frac{2x^2-1}{\sqrt{x}}\, dx$

(E.M.E.U.)

3. Show that the curve $y = 4 - x^2$ cuts the x-axis at $x = -2$ and $x = 2$. Sketch the curve, and find the area of it above the x-axis. (U.E.I.)

4. Find the area between the graph of $y = (4 - x)(x + 2)$, the x-axis and the ordinates at $x = 0$ and $x = 3$,
   (i) by Simpson's rule with 6 strips
   (ii) by direct integration. (N.C.T.E.C.)

5. Plot the graph of $y = x(6 - x)$ for values of x from 0 to +6. Use Simpson's Rule to estimate the area enclosed between the curve and the x-axis. Hence, or otherwise, evaluate

$$\int_0^6 x(6 - x)\, dx$$ (U.E.I.)

6. (a) Integrate with respect to x:

   $y = (4x+3)^{-2}$

   (b) Use Simpson's rule with six strips to obtain an approximate value of the integral $\displaystyle\int_0^6 \frac{1}{1+x^2}\, dx$ (N.C.T.E.C.)

7. (a) Integrate the following expressions:

   (i) $(x^4-3x^2-5)\, dx$ (ii) $54t(3t^2-2)^8\, dt$ (iii) $\dfrac{3}{x^3}\, dx$

   (b) The curve $y = \sqrt{9-x^2}$ is revolved around the x axis through an angle of 360°. Calculate the volume bounded by this curve between the points $x = 0$ and $x = 3$. (U.E.I.)

8. (a) Obtain the following integrals:

   (i) $\displaystyle\int \frac{x^3 - 1}{x^2}\, dx$ (ii) $\displaystyle\int 4\sqrt{x}\, dx$

   (b) The gradient at any point (x, y) on a certain curve is given by the expression $3x^2 - 4x + 3$. Given that the curve passes through the point (2, 5) find its equation.

   (c) Given that $dx/dt = u + at$ (where u and a are constants), find a formula for x in terms of u, a, and t, given that $x = 0$ when $t = 0$. (U.E.I.)

9. A cantilever of length l m and weight w newtons per metre has one end clamped horizontally and carried a load of W newtons at the other. At a point x m from the fixed end the deflection is y m below the fixed end and these variables are related by the formula

$$EI\frac{d^2y}{dx^2} = \frac{w}{2}(l-x)^2+W(l-x)$$

Show that the deflection at the free end is equal to

$$\frac{l^3}{24EI}(3wl+8W) \qquad \text{(W.J.E.C.)}$$

10. (a) The gradient at a point (x, y) on a given curve is $3x^2-18x+24$. If the curve has an intercept 12 on the y axis, obtain its equation.

(b) For a beam of length L, clamped at one end ($x = 0$, $y = 0$) and freely supported at the same level at the other end ($x = L$, $y = 0$), with a uniformly distributed load over the whole span of w per unit length,

$$EI\frac{d^2y}{dx^2} = \tfrac{1}{8}wL^2-\tfrac{5}{8}wLx+\tfrac{1}{2}wx^2$$

Find the deflection y, in terms of x and the constants E, I, W and L, and prove that the maximum deflection occurs when

$$x = \frac{L}{16}(15-\sqrt{33}) \qquad \text{(U.L.C.I.)}$$

12 Probability

WHAT do we mean by expressions such as ' nine times out of ten ' and ' odds of four to one '? How does an insurance firm calculate its premiums? How can a builder estimate the chance of wet weather delaying certain stages of a construction project?

All these are aspects of probability: i.e. the branch of mathematics which enables us to calculate the likelihood of any particular outcome.

Basic definitions

In any problem or experiment, each separate result is called an **outcome.** The particular happening we are looking for will be called the **event.** We shall define the **probability** of the event by $p = r/n$ where n is the number of observations during which the event was the outcome on r occasions.

Two events are **complementary** when together they include every possible outcome without any overlap or duplication. When all possible outcomes are known, the set of outcomes can be referred to as the **sample space.**

Example When a couple have their first child, obviously it must be either male or female. Hence, for one child, the complete set of possible outcomes is {M, F}.

Assuming equal probability for each possible outcome, the probability of the first child being male is $\frac{1}{2}$ and the probability of it being female is equally $\frac{1}{2}$. Note that the probabilities of the complementary events add together to give 1.

Now consider a family with two children. The sample space is {MM, MF, FM, FF}. By symmetry, it is obvious that each of these possible outcomes, being equally likely, have a probability of $\frac{1}{4}$ (i.e. 1 in 4).

If we consider the possibility of such a family with two children having at least one boy, we can either add the individual probabilities for MM, MF, FM to get $\frac{1}{4} + \frac{1}{4} + \frac{1}{4} = \frac{3}{4}$, or we can use our basic definition from which $r = 3$, $n = 4$, and thus $p = \frac{3}{4}$.

Non-occurrence

If the probability that a certain event may happen is p, then the probability that the event will not happen is $1 - p$.

A probability of zero implies that there should be no chance at all of the event occurring.

Note that negative probabilities are impossible, and that p can never be greater than 1 since $p = 1$ implies a certainty.

Example: The probability of having a year in England during which there is no rainfall is almost certainly zero and, conversely, the probability is therefore 1 that it will rain at least once.

Equal probabilities

The first example considered above assumed equal probabilities in respect of male or female births and this assumption is fairly reasonable in respect of human births (though not, say, for honey bees!). Many of the problems on which we can do experiments are of this type where all possibilities are equally likely, and the principles are the same whether the experiments are concerned with tossing coins, rolling dice, or selecting cards. It is important to distinguish between problems involving equal probabilities and other types of problems.

Assuming we have dice which are properly balanced, when a single die is rolled, it may come to rest with any of its six faces uppermost. The probability of any particular face appearing is thus $\frac{1}{6}$. However, if we attempted to roll an ordinary brick instead, it is obvious that it is far more likely to come to rest on either its top or bottom face than to stand on one end! In the case of a rolling brick, therefore, the probabilities are clearly unequal, and the probability of the brick coming to rest on one particular end is obviously less than $\frac{1}{6}$.

Unequal probabilities

Most questions of this type are real problems arising in situations where evidence has to be collected to determine experimentally the particular probability. For example, a football team has a higher probability of winning at home than of winning the corresponding away match.

Many of the problems in this category can be approached by using the statistical techniques outlined in the next chapter on the basis of actual experiments and measurements; however, some problems, which appear to involve unequal probabilities at a first glance, turn out to be based on groups of outcomes where the outcomes may be equally probable but the groups of different size. To put it in the language of modern mathematics, we may have a sample space in which each individual outcome is equally probable, but we can define sub-sets of different sizes and thus the probabilities of the sub-sets can be different.

Example: Consider the sample space for a pair of dice. (If we label the first die as x and the second die as y, we can use the analogy of Cartesian co-ordinates to produce an ordered array which will simplify subsequent references.)

$$\begin{Bmatrix} (1,6) & (2,6) & (3,6) & (4,6) & (5,6) & (6,6) \\ (1,5) & (2,5) & (3,5) & (4,5) & (5,5) & (6,5) \\ (1,4) & (2,4) & (3,4) & (4,4) & (5,4) & (6,4) \\ (1,3) & (2,3) & (3,3) & (4,3) & (5,3) & (6,3) \\ (1,2) & (2,2) & (3,2) & (4,2) & (5,2) & (6,2) \\ (1,1) & (2,1) & (3,1) & (4,1) & (5,1) & (6,1) \end{Bmatrix}$$

Now our sample space contains 36 possible outcomes, and since each individual outcome is equally probable this implies a probability for a single outcome of $\frac{1}{36}$. Should we wish to know the probability of throwing a *double* with our two dice, then we are really asking for the probability of the sub-set of outcomes along the diagonal ($x = y$). This subset contains six possible outcomes out of 36 and its probability is therefore $\frac{6}{36}$ or $\frac{1}{6}$. We can compare this with the probability for any other subset: e.g. the probability of scoring a total of 3 or less with the two dice is $\frac{1}{12}$ and, conversely, the probability of scoring 4 or more is $\frac{11}{12}$.

Events in sequence

When we come to consider probabilities for events happening one after the other, there are four basic ways in which events may (or may not) be related. Let us consider each of these in turn, because if we can distinguish clearly between them it will save considerable confusion.

Independent events

It is important to distinguish between events which are independent and those which are not. We shall define independent events as those which can happen (or not happen) without affecting the probabilities of the other events in the sequence.

The probability that two independent events will both happen is given by the *product* of their individual probabilities: i.e. $p = p_1 \times p_2$. This principle can be extended to cover any number of independent events.

Example 1 If a coin is tossed, it may come down showing the head side or the tail side. Since each outcome is equally likely, the probability of either result is $\frac{1}{2}$. When the coin is tossed again, the probabilities are still exactly the same and are unaffected by the first result. Hence the probability of a sequence of two heads occurring in the two attempts is given by

$$p = \tfrac{1}{2} \times \tfrac{1}{2} = \tfrac{1}{4}$$

Similarly, the probability of a sequence of three heads in three attempts is

$$p = (\tfrac{1}{2})^3 = \tfrac{1}{8}$$

The argument can be extended to show that a sequence of n heads in n attempts will have a probability of $(\frac{1}{2})^n$.

Example 2 Taking a standard set of 52 playing cards, shuffling them thoroughly, and selecting a single card at random should give a probability of selecting an ace of $\frac{1}{13}$ since there are four aces distributing amongst the total of 52 cards and $\frac{4}{52} = \frac{1}{13}$. *If this card is now replaced* and the pack reshuffled, the probability of selecting an ace at the second attempt is also $\frac{1}{13}$ and the probability of both choices being aces is

$$\frac{1}{13} \times \frac{1}{13} = \frac{1}{169}$$

Dependent events

Whenever a particular choice affects subsequent choices, the events are dependent.

In the previous example, if an ace is selected from the set of 52 cards on the first choice and this card is *not* replaced before the second choice is made, trying to select a second ace from the remaining 51 cards means picking one of the three aces which are still left in the pack. Hence the probability of the second choice *also* being an ace is reduced to $\frac{3}{51} = \frac{1}{17}$. Therefore, *without replacement*, the probability of picking two aces, one after the other, is given by

$$\frac{1}{13} \times \frac{1}{17} = \frac{1}{221}.$$

Continuing the argument gives a probability of 1/5525 for picking out three aces one after another.

EXERCISE 12(a)

1. If the probability that Manchester United will win a football match played on their home ground is 0.6, why is the probability that they might lose the home match less than 0.4?

2. In a slot machine there is an arrow which spins round horizontally in such a way that it is equally likely to stop at any of the sectors numbered from 1 to 32 round the dial. Calculate the probabilities in each case for the arrow to stop:
   (a) on an even number,
   (b) on a number ending in 5,
   (c) on a number divisible by 3.

3. If a pack of 52 cards is shuffled and then four cards are selected (without replacement, find the probabilities for the following:
   (a) all four cards were Queens,
   (b) all four cards were diamonds,
   (c) all four were face cards (i.e. Jack, Queen, or King).

4. Consider the sample space of Cartesian coordinates such that both x and y are positive whole numbers defined by $0 \leqslant x \leqslant 4$ and $0 \leqslant y \leqslant 4$, write down the probabilities for:
   (a) a point selected at random from this set having coordinates (2, 3),
   (b) a point being on the x-axis,
   (c) a point having x and y coordinates which are equal.

5. A normal coin is tossed three times and the results recorded as heads, heads, tails. What is the probability that the outcome of the next toss will be (a) heads, (b) tails?

6. A box contains two dozen hexagonal nuts which are the same external size and all look alike, but four of them have the wrong internal thread to fit the bolts which are available. If three bolts have to be paired up with nuts, find the chance that the first three nuts selected will all be of the wrong thread.

7. The probability of player X winning a round of golf against player Y is $\frac{1}{4}$. Show that the probability of X winning at least one round out of three against Y is $\frac{37}{64}$.

8. A student who wishes to study part-time for two subjects at pre-university level finds that the local College offers a choice of five science subjects and six arts subjects. Calculate the number of ways in which he can choose one science and one arts subject.

9. Calculate how many three-letter combinations can be made with the 26 letters of the alphabet:
   (a) if each letter is not used more than once,
   (b) if letters may be repeated.

10. If a certain political party actually has the support of 20% of the voters, find the probability of four voters, selected at random, all being supporters of this particular party.

Dependent or independent events

In the first section of this chapter we have investigated how to calculate the probability of a single event happening. We have then considered how to find the probability of two or more events happening together or in sequence. In this next section we shall consider how to calculate the probabilities of *either* one *or* another event occurring.

Mutually exclusive events

This is the easier case where the alternatives do not overlap at all and the basic rule is simply to add all the individual probabilities:

$$p = p_1 + p_2$$

In fig. 48(a), S represents the sample space and A and B are subsets of possible outcomes which do not overlap.

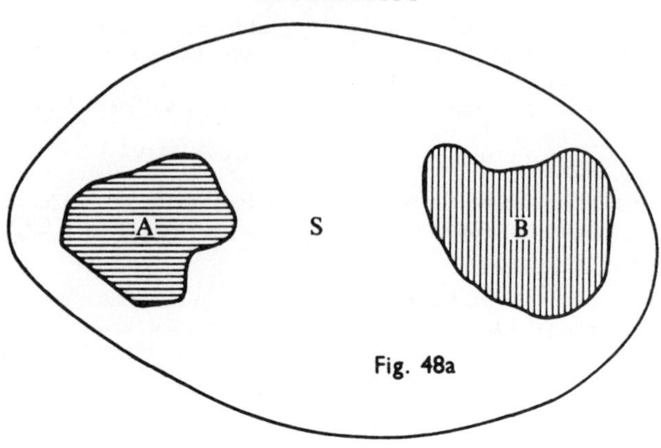

Fig. 48a

The probability of an event being in S is 1,
the probability of it being in A is $p(A)$,
the probability of it being in B is $p(B)$,
so the probability of it being in A *or* B is $p(A) + p(B)$
and it should be obvious that $p(A) + p(B) \leqslant 1$.

Either or both

Figure 48(b) represents diagrammatically the situation when subsets A and B overlap. In this case it is clearly possible for an event to be in A, or to be in B, or to be in *both A and B*. This duplication means that simple addition of probabilities will not be correct since it implies that the area shared by A and B will be counted twice.

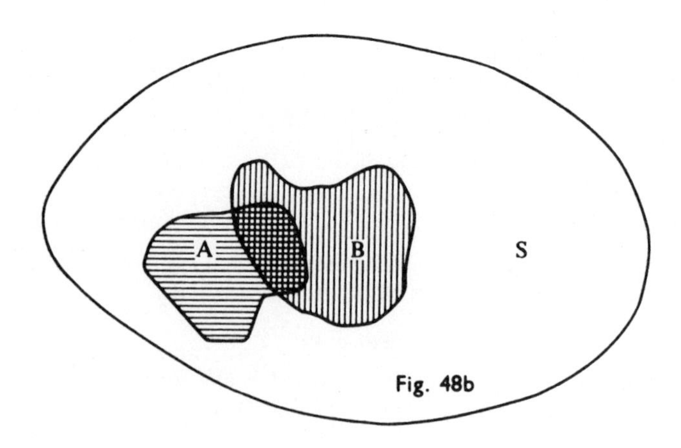

Fig. 48b

Example Let us take for our sample space S the one we considered previously for rolling two dice x and y. If we wish to find the chance of scoring a total *either* 4 or less *or* 10 or more, we can consider the subset of outcomes where the score is 4 or less to be subset A and we can label the subset where the score is 10 or more as subset B (fig. 49). From this diagram it is easy to see that the two possibilities are mutually exclusive (which is what we would expect since the score must be either high or low but cannot be both at the same time!

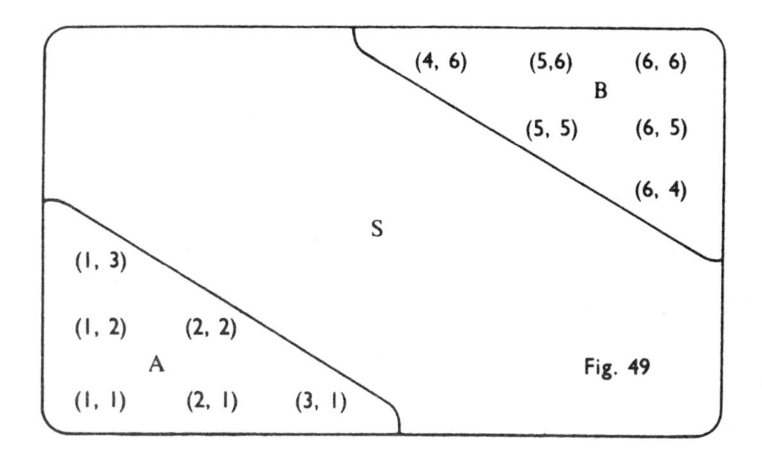

Fig. 49

Subset A contains 6 possible outcomes out of the sample space of 36, $\therefore\ p(\mathrm{A}) = \frac{1}{6}$. Similarly, $p(\mathrm{B}) = \frac{1}{6}$ and hence the chance of scoring either 4 or less or 10 or more is given by

$$p = p(\mathrm{A}) + p(\mathrm{B}) = \tfrac{1}{3}$$

As a simple example of events which are *not* mutually exclusive, consider the probability for at least one ⚂ occurring when we toss the two dice. Here the two subsets overlap at the point (3, 3) as shown in fig. 50. Having defined $p(\mathrm{A})$ as the probability of an event in A, and $p(\mathrm{B})$ as the probability of it occurring in B, we must now define $p(\mathrm{AB})$ as the probability of the event being in the area shared by both A and B. Since $p(\mathrm{A}) + p(\mathrm{B})$ includes $p(\mathrm{AB})$ *twice*, and this possibility should only be included once, our more general form of expression for the probability of an event being in either A or B or both becomes

$$p = p(\mathrm{A}) + p(\mathrm{B}) - p(\mathrm{AB})$$

For those familiar with the notation of modern mathematics this is more clearly expressed as

$$p(\mathrm{A} \cup B) = p(\mathrm{A}) + p(\mathrm{B}) - p(\mathrm{A} \cap \mathrm{B})$$

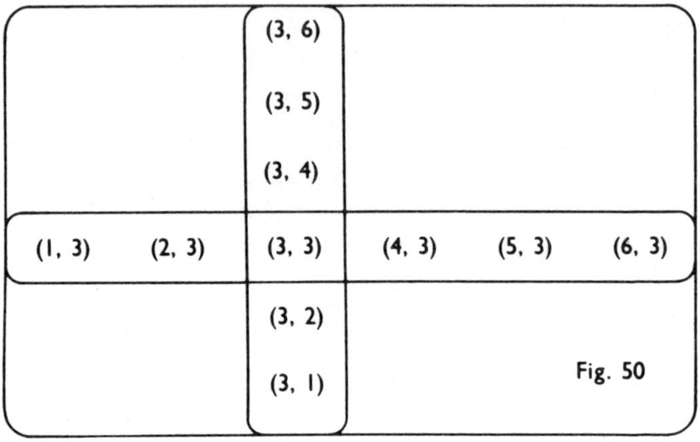

(3, 6)

(3, 5)

(3, 4)

(1, 3) (2, 3) (3, 3) (4, 3) (5, 3) (6, 3)

(3, 2)

(3, 1)

Fig. 50

The extension to three or more events which are not mutually exclusive is beyond the scope of this book but can be given as

$$p = p(\text{A}) + p(\text{B}) + p(\text{C}) - p(\text{AB}) - p(\text{BC}) - p(\text{CA}) + p(\text{ABC})$$

or in the alternative form as

$$p(\text{A} \cup \text{B} \cup \text{C}) = p(\text{A}) + p(\text{B}) + p(\text{C}) - p(\text{A} \cap \text{B}) - p(\text{B} \cap \text{C}) - p(\text{C} \cap \text{A}) + p(\text{A} \cap \text{B} \cap \text{C})$$

Using this formula you can calculate that the chance of throwing at least one six when throwing three dice together is 91/216, but it must be realised that this is the theoretical probability, and it does not necessarily follow that if you toss three dice 216 times that you will get at least one six on exactly 91 occasions. You could try this experiment several times and see how closely your experimental results agree with the theoretical prediction.

Example A small building firm analyses a working year averaging 50 weeks per employee and finds that in this period 3 weeks were lost due to industrial disputes and 5 weeks due to bad weather. Calculate the probability for any average week that

(a) The men will be on strike.
(b) The men will be rained off.
(c) The men will be working normally.

(a) Probability of industrial dispute $\frac{3}{50} = 0.06$
(b) Probability of bad weather $\frac{5}{50} = \frac{1}{10} = 0.10$
(c) Probability of both together is the product $= 0.006$
$\therefore$ probability of not working $= 0.06 + 0.10 - 0.006 = 0.154$
$\therefore$ probability of normal working is $1 - 0.154 = \underline{0.846}$

EXERCISE 12(b)

1. A coin is tossed and a single die rolled. Determine the probability:
   (a) that the coin shows heads *and* the die shows 6,
   (b) that either the coin shows heads *or* the die shows 6.

2. A medical doctor estimates that 1 in 10 of the patients he sees have an infection and that 1 in 8 of his patients need sleeping pills. What is the probability that his next patient will have an infection *and* need sleeping pills?

3. 25 raffle tickets are sold and there are two prizes. Calculate the probability that a certain man who has bought two of the tickets wins (a) neither prize, (b) both prizes, (c) either one or the other prize.

4. A man estimates that the probability of him surviving for the next 20 years is 0.20, but the probability that his wife will still be alive after 20 years is 0.30. Calculate the probabilities that after 20 years:
   (a) both will still be alive,
   (b) neither will be alive,
   (c) only the wife will survive.

5. If 5% of the concrete tested on a particular contract is below the specified acceptable strength, and if 10% of the reinforcing rods are faulty, estimate the probability that a particular sample of the resulting reinforced concrete will have *either* weak concrete *or* faulty reinforcing rods.

6. The clerk of works reports on a group of 24 houses that 4 had plumbing faults, 6 had joinery faults and 3 had electrical faults. Calculate the probability that a particular house
   (a) had all three types of fault,
   (b) had none of these faults.

REVISION EXERCISE 12

1. Out of 40 paving slabs, 12 have internal cracks. Determine the probability that, in choosing 5 slabs at random, you avoid selecting one which is cracked.

2. Estimate the probabilities that any two people selected at random will have their birthdays:
   (a) in the same month,
   (b) on exactly the same date.
   Why is it not possible to give these two probability figures exactly?

3. The College Refectory presents a choice of soup or fruit juice for the first course, a selection of six different main courses and a choice of four sweets. How many different possible combinations exist for a three-course lunch?

4. A contractor has to excavate for foundations in an area for which there are no available plans showing the position of existing services.
   It is anticipated that he may cut across one or more service lines whilst excavating and on the known areas probabilities are calculated for each of the following services:
   (1) Water supply, 0.08;
   (2) Drainage, 0.12;
   (3) Gas, 0.05;

(4) Electricity, 0.05;
(5) Telephone cable, 0.04.

Estimate the probability of cutting across
(a) one or more of these services,
(b) both water supply *and* drainage,
(c) either electricity *or* telephone cable.

5. A builder has acquired a large quantity of nuts and bolts of a certain standard size, but some of them are faulty. If 4% of the nuts are faulty and 5% of the bolts are faulty, find the probability that if any bolt is selected and paired with any nut chosen at random,
(a) only the nut will be faulty,
(b) only the bolt will be faulty,
(c) both of them will be faulty,
(d) at least one of them will be faulty,
(e) neither of them will be faulty. (U.E.I.)

6. If you were one of six men on a building site and you all agreed to roll a die to decide who should make the tea at the morning break,
(a) What would be the probability that your number would show on the die on both the first and the second occasion?
(b) Prove that the probability that you would have to make the tea on either the first or the second occasion or both would be less than $\frac{1}{3}$.
(c) Show that the probability is greater than $\frac{1}{2}$ that you would have to make the tea at least once for the first four times. (U.E.I.)

7. An analysis of 20 O.N.C. students in the Building Department of a certain College showed the following distribution:

Age	17	18	19	20	Total
Male	4	6	3	3	16
Female	1	2	1	0	4
Total	5	8	4	3	20

If four of the enrolment cards of these students are selected at random, calculate the probability for each of the following.
(a) The cards relate to students who are all age 17.
(b) The four students are all of different ages.
(c) The four are all male students.
(d) None of the four are 18. (U.E.I.)

8. A man has in his drawer six identical brown socks and ten similar blue socks. The room is in total darkness and he goes to the drawer to get a pair of socks.
(a) Find the minimum number of socks he should pick up out of the drawer and take to the nearest light in order to ensure:
(i) that he has at least two socks which match;
(ii) that he has a brown pair;
(iii) that he has a blue pair.

(b) Calculate the probability that if he selected any two socks at random, he would find subsequently that he had:
  (i) two odd socks,
  (ii) a brown pair,
  (iii) a blue pair.

9. What is the probability of getting at least one six in two throws of a single die?

10. In a raffle, 100 blue tickets are sold and 150 red ones. If a man buys 2 blue tickets and 5 red ones what is his chance of winning the first prize?

11. The probability that a certain man will be alive in 20 years time is 0.60 and the probability that his wife will be alive in 20 years time is 0.75. Calculate the probabilities that in 20 years time
(a) both will be alive,
(b) neither will be alive,
(c) at least one will be alive.

12. Sixteen men were at work demolishing an old factory when part of it collapsed, unexpectedly injuring three of them. If three of the men were apprentices, calculate the probability that all three were injured.

13. 10 paving slabs are delivered to a site and, although they all look alike, there are 2 mixed in the delivery which are substandard and liable to crumble in heavy frost. If only 6 of the slabs are laid, calculate the probability that neither of the substandard ones were actually used.

14. If it is normal to expect rain in 10 days out of the 28 in the month of February, what chance is there that a particular period of 7 days will be completely dry.

13 Statistics

THE increasing use of statistics in the building industry is leading to considerable economy and greater reliability in many ways. In combination with efficient recording and effective programming the application of statistical methods has enabled remarkable results to be achieved by certain firms and many others are now following their example.

Application to strength of materials

Most of the materials commonly used in building show a considerable variation in strength, and this variation is usually covered by the use of a comparatively large safety factor. If, for example, we select forty test pieces of a typical softwood, taking care to avoid all visible defects, then ensure that they are all the same size and have an equal moisture content and are thus seemingly identical, nevertheless it would be quite possible for these forty specimens to show a range of strength from 28 to 56 N/mm² when subjected to crushing tests in the laboratory. The average for the forty test pieces may be, say, 42 N/mm², in which case the variation is up to 33⅓% either side. Now by applying statistical analysis to the figures obtained from our tests we may estimate not only the average strength and the extent of the variations from it, but also the probability of finding among a further group of specimens one with a strength even lower than 28 N/mm² and this is obviously of vital importance in the safe and yet economical use of this particular material.

In a prefabricated material such as a concrete block, statistical analysis may be applied in experimental studies to determine the most economical composition yielding the required strength together with consistent reliability.

We may also measure statistically the advantages of laminated construction, and investigate the properties and reliability of the many new prefabricated materials now being produced experimentally.

Application to programming, etc.

Large variations exist in the labour expenditure by various contractors on similar contracts. These may be analysed into differences caused by variation in the amount of experience of personnel, the extent of subcontracting, size of contract, region, bonus incentive schemes and influence of the principal of the firm. Statistical investigation can reveal how to carry out the programming of the work on a contract so that it may be executed as quickly and as efficiently as possible. The extent of the variation is shown by the fact, that for a traditional medium-sized house, on a contract for 20 to 30 such houses the average labour expenditure by skilled bricklayers is approximately 600 hours, but about one-sixth of the contractors can complete the bricklaying with an expenditure of 400 hours, or less, whilst one-sixth actually take more than 800 hours exclusive of travelling time and time lost due to bad weather. Generally speaking, firms which operate bonus incentive schemes, or those in which the principal is actually working on the site, average some 100 hours less than the remainder. This shows that there is considerable scope for increased productivity in the building industry as a whole.

Recording the information

The actual figures may often be recorded in tables in order, but, if it is desired to enable an interested person to receive at a glance an impression of the significance of the figures, it is useful to use some mode of pictorial representation. Some examples are given below.

Other trades 5.2 %
Plumbers 6 %
Painters 7.4 %
Plasterers 11.1 %
Carpenters 14.8 %
Bricklayers 22.2%
Site Works (labourers) 33.3%

Fig. 51
The 100% Bar Chart

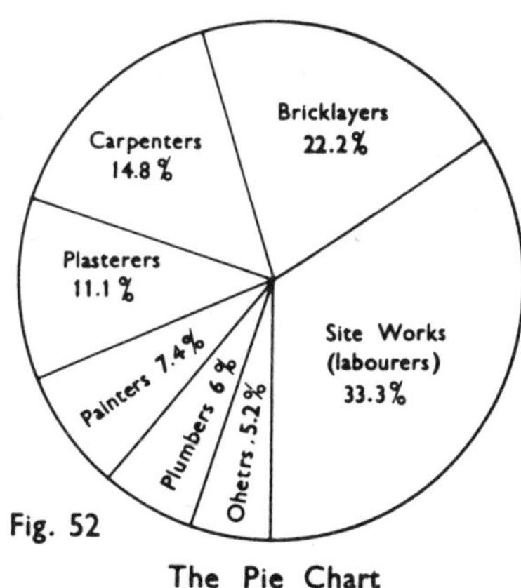

Fig. 52
The Pie Chart

Figure 51, the 100% Bar Chart, and fig. 52, the Pie Chart, show the distribution of labour in the building of one house of traditional design; the figures include work subcontracted. In each of these diagrams the proportion of work for any trade is represented by an area (which may be left blank or suitably shaded or coloured) and either form of presentation enables comparisons to be made at a glance.

Figure 53, the Horizontal Bar Chart, is based on the same information, but in this case the actual man-hours are shown, so that this type of diagram conveys more information. It compares the time for one trade with that for the others rather than with the whole time taken. For quantities which vary discontinuously with time a vertical bar chart may be used, e.g. to illustrate the rainfall figures for each month of a particular year.

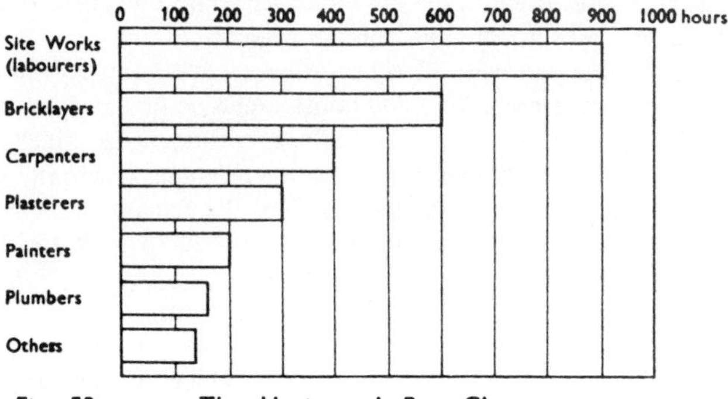

Fig. 53 The Horizontal Bar Chart

The most widely used method of depicting information is the graph. When the points have been plotted from the information supplied they may be connected by joining them with straight lines, or a smooth curve may be drawn as nearly as possible through the points. Straight lines should be used when there is no definite law; the curve indicates that intermediate values are reasonably reliable. A familiar example of straight line connection is the temperature chart of a hospital patient.

Frequency distribution table

If a survey were to be made of the number of electricity power points in 75 houses chosen at random, the results of the survey might be recorded as follows:—

Number of points	Houses with this number	
2	/	1
3	///	3
4	~~////~~ //	7
5	~~////~~ ~~////~~ ~~////~~	15
6	~~////~~ ~~////~~ ~~////~~ ~~////~~	20
7	~~////~~ ~~////~~ ///	13
8	~~////~~ ///	8
9	////	4
10	//	2
11	/	1
12	/	1
	Total	75

The first column gives the values of x, the variate: the last column gives the values of f, the frequency, showing how many times each particular value of x occurs.

Three methods of depicting the relationship between variate and frequency are illustrated below.

The histogram

This form of diagram (fig. 54) provides an immediate visual comparison of the respective frequencies. This comparison is really based on the relative *areas* of the columns to allow for cases where the base line cannot conveniently be split up into equal intervals, but in the great majority of cases

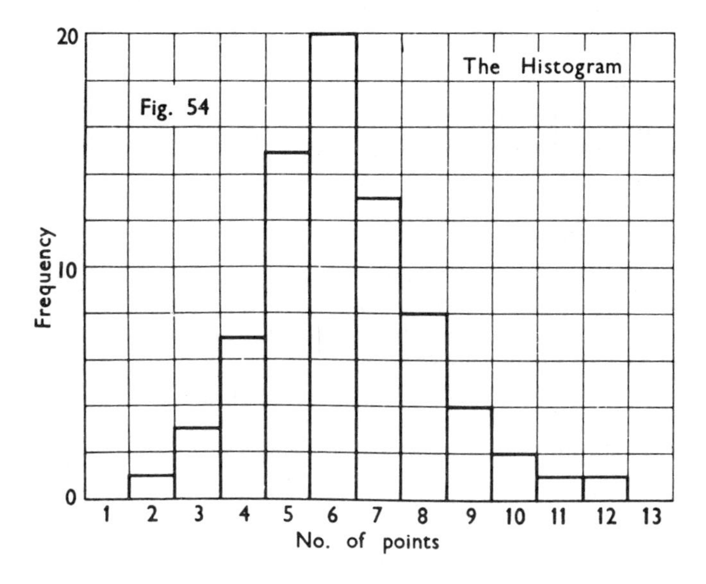

the intervals are made equal, and then the length of each column is proportional to the frequency. If the intervals are not equal, no vertical scale should be shown. The total frequency of the distribution is represented by the total area of the figure. For a well-balanced figure, the greatest height should be roughly $\frac{2}{3}$ to $\frac{3}{4}$ of the width.

A certain examination was taken by 150 students and the following table shows the distribution of the percentage marks obtained.

% Marks	0 → 25	→ 35	→ 40	→ 45	→ 50	→ 60	→ 70	→ 80	→ 100
Frequency	5	12	16	25	23	29	18	14	8

Note that in this case the percentage marks are given in class intervals which are unequal. The first interval covers the range 0 to 25, but the second and third intervals are for successively shorter ranges. The histogram for these results is therefore constructed as shown in fig. 55.

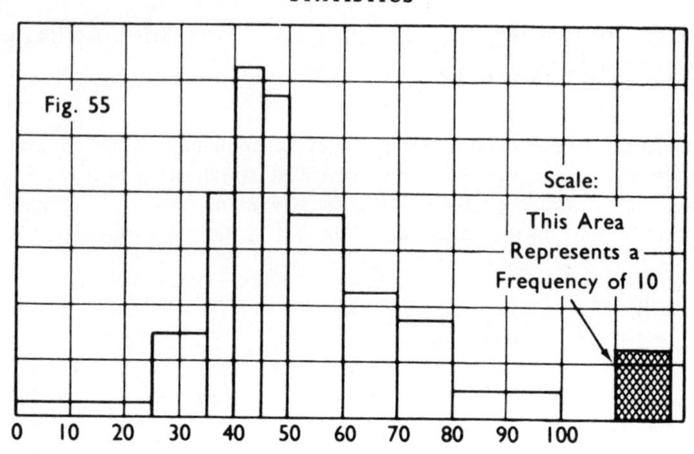

The frequency polygon

This is an alternative to the histogram, but is only to be preferred when one distribution is superimposed upon another for purposes of comparison.

With the frequency polygon (fig. 56) the intervals along the axes *must* be equal, and an additional value (of zero frequency) should be included at each end of the given range of the distribution in order that the area enclosed by the polygon may represent the total frequency. The frequency polygon is sometimes superimposed on the histogram simply by joining the mid-points

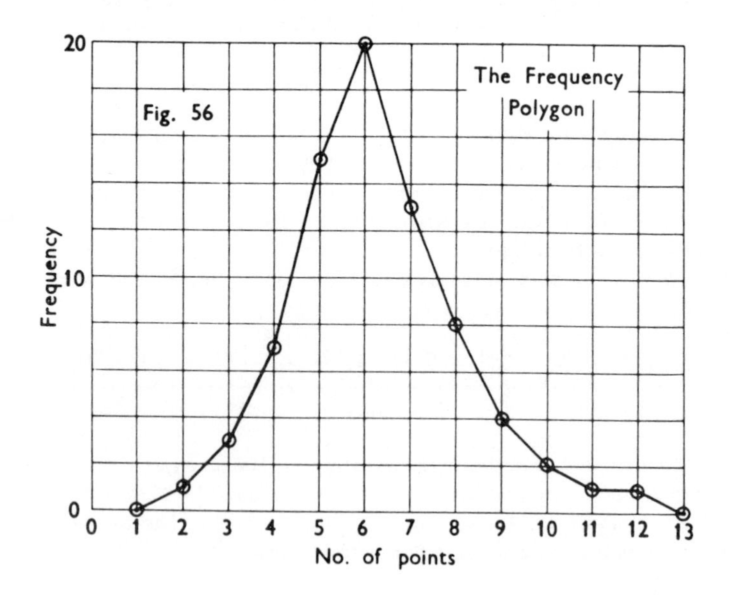

of the tops of the columns: there is no point in doing this, and the practice is definitely not recommended.

The distribution curve

When the figures in the frequency distribution table represent only a sample, it is reasonable to assume that if more such samples were taken the figures would be similar but not exactly the same. The net effect of combining the frequency polygons for a whole set of samples would be to smooth out the outline and the result would be a curve (fig. 57). Certain types of distributions correspond to the ' normal ' distribution curve, which is a symmetrical peaked curve extending indefinitely in both directions close to the horizontal axis. The true normal curve is rare, but many distributions conform to it approximately. It can be defined mathematically in terms of the total number of observations, the arithmetic mean and the standard deviation (see later).

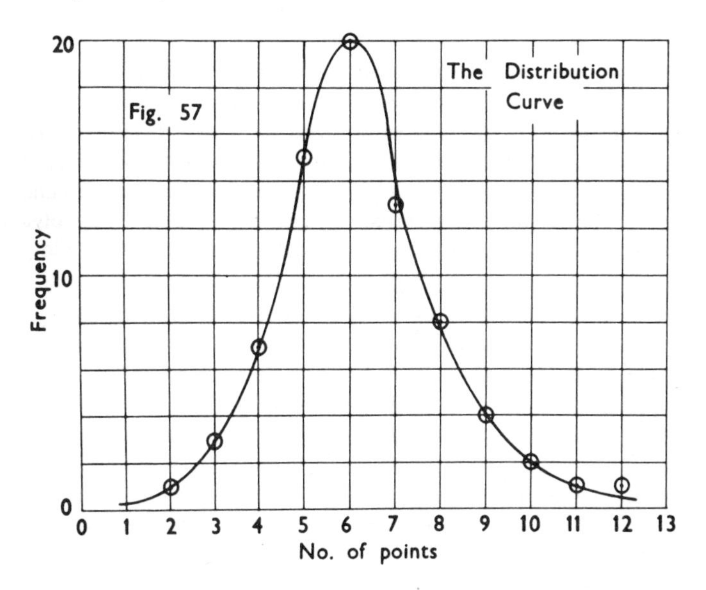

Types of average

(1) The **arithmetic mean**—commonly referred to as the A.M. or simply, the *mean*. This is the average most frequently in use and most widely understood. It is obtained by adding together all the values of the items in the distribution and dividing the total by the number of items. This is conveniently expressed as

$$\bar{x} = \frac{1}{n}\sum x$$

where $\bar{x}$ is the mean, n is the number of items in the distribution, $\sum$ is a mathematical symbol meaning " the sum of all the values of " and x is the variate.

For a grouped distribution each value of the variate, x, must be multiplied by the corresponding frequency, f, and in this case the mean is given by

$$\bar{x} = \frac{1}{n} \sum fx$$

(2) The **geometric mean**—the G.M. is the nth root of the product of the n items.

$$G = \sqrt[n]{x_1 . x_2 . x_3 \ldots x_n}$$

thus $$\log G = \frac{1}{n} \sum \log x$$

and for a grouped distribution:

$$\log G = \frac{1}{n} \sum \mathrm{f} \log x$$

(3) **The median**—the central value. To find the median, the complete distribution must be arranged in order of size; then, with an odd number of items, the central value will be the median: with an even number of items the median is the average of the central pair. With a grouped distribution, first pick out the group containing the median by summing the frequency column as far as the value $n/2$ and then evaluate the median by proportion. (For example see later.)

(4) The **mode**—the most frequently occurring value. The mode corresponds to the peak of the distribution curve. It is the variate with the greatest frequency. For a grouped distribution the **modal class** can be defined as the class with the greatest frequency.

Each of these averages possesses certain advantages and they are used in different circumstances, the aim being to select the particular average which is representative of the whole distribution. It should be noted that the mean is unduly affected by the existence of extreme items and is unlikely to coincide with any of the original values, e.g. ' each family had an average of 2¼ children '! The mean has the advantage of being well known and, furthermore, the total of all the items in the distribution may be regained from it, as

$$\sum x = n\bar{x}$$

Except in the case of a grouped distribution the median and the mode are actually occurring values; they take no account of extreme items, but may tend to give a false impression with a *skew distribution*, i.e. one in which the peak of the distribution curve is displaced to one side of the centre of the distribution. The geometric mean is used in connection with index numbers (which are not dealt with in this volume): its main disadvantage is that it cannot be used if any item is zero or negative. The arithmetic mean is used in the analysis of experimental results when they are reasonably consistent, and is the basis of many further statistical calculations. The median and the mode are used when an actual representative value is required.

Moving average

This is an extremely useful method of estimating the trend of a time series. If, for instance, monthly production figures are plotted graphically, the line of the graph will probably fluctuate considerably due to seasonal variation; the moving average is a method of smoothing out these fluctuations to reveal the trend and enable forecasts of future production to be made as accurately as possible. For monthly figures with a yearly cycle of seasonal variation the twelve-monthly moving average would be required. For figures covering several years, four-quarterly moving averages would be more convenient, while for much shorter periods of a few weeks, seven-day moving averages would be advisable.

Example: The following table gives day attendance figures at a certain small theatre for each quarter over a period of three years. Plot these values and superimpose the graph of the quarterly moving average. It is anticipated that if the present trend continues the theatre may have to be closed when the average attendance drops below 600 per week. Estimate when this is likely to happen.

	Year I	Year II	Year III
Jan–Mar	12 400	11 300	10 300
Apr–Jun	11 500	10 600	9 500
Jul–Sep	11 300	10 300	9 200
Oct–Dec	11 100	9 900	8 900

The first average is for the four quarters of Year I:

$$a_1 = \tfrac{1}{4}(12\,400 + 11\,500 + 11\,300 + 11\,100) = 11\,575$$

On the graph this value is plotted at the centre of the Year I interval.

The second average is for the last three quarters of Year I together with the first quarter of Year II. It may be calculated as

$$a_2 = \tfrac{1}{4}(11\,500 + 11\,300 + 11\,100 + 11\,300) = 11\,300$$

but it is usually more convenient to work from the previous average by adding the appropriate fraction for the difference between the last figure included and the one displaced, thus:

$$a_2 = a_1 + \tfrac{1}{4}(11\,300 - 12\,400) = 11\,575 - 275 = 11\,300$$

On the graph this value is plotted opposite the end of the third quarter of Year I.

Set out below is the complete table of the given figures together with the calculated moving averages.

	Year I		Year II		Year III	
Jan–Mar	12 400		11 300		10 300	
				10 825		**9 725**
Apr–Jun	11 500		10 600		9 500	
		11 575		**10 525**		**9 475**
Jul–Sep	11 300		10 300		9 200	
		11 300		**10 275**		
Oct–Dec	11 100		9 900		8 900	
		11 075		**10 000**		

The graph is easily plotted with the moving average superimposed. An average attendance of 600 per week would be 7800 per quarter and it is evident from the graph of the moving average that if the present downward trend continues at the same rate the theatre will have to be closed sometime during the first quarter of Year V, although a margin of approximately three months either way should be allowed to cover fluctuations in the actual attendance figures.

It should be noted that the graph of the moving average is not necessarily a straight line and it is therefore not always easy to obtain a reliable forecast. Should the graph of the moving average show considerable fluctuations this indicates that the time interval selected was too short.

The cumulative frequency curve

Another way of showing a distribution is by a cumulative frequency curve or **ogive.** This curve illustrates how the total frequency is built up, and is useful in that it facilitates estimates of the proportion of the distribution between set limits. To plot the ogive, the table of values must be extended to include another set of figures. These can be the cumulative frequencies, obtained by adding together the preceding frequencies, or alternatively, it is sometimes an advantage to express the cumulative frequencies as percentages.

Example: Absorption tests on 100 standard concrete blocks gave the following results:

Absorption below	8%	9%	10%	11%	12%	13%	14%	15%	16%
No. of blocks	1	2	8	22	35	21	7	3	1
Cumulative frequency	1	3	11	33	68	89	96	99	100

From the curve in fig. 58 it is easy to see, for example, that a third of the concrete blocks have an absorption of not more than 11%.

Lines drawn to intersect the cumulative frequency curve at the 25%, 50% and 75% levels divide the distribution into four sections known as **quartiles.** The central (50%) line is the median of the distribution (in this case 11.5%

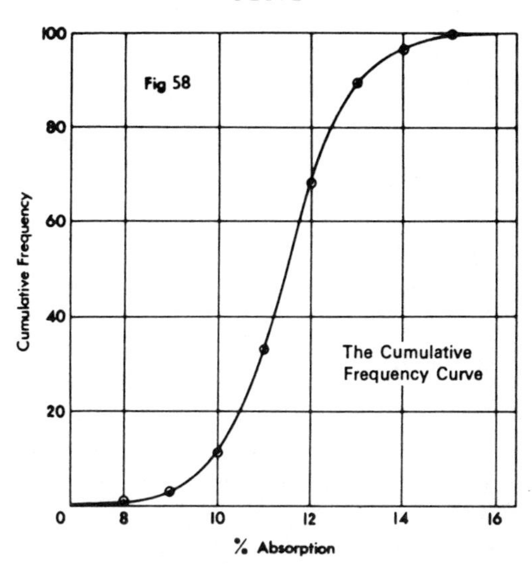

absorption). The upper and lower quartiles are at 12.2% and 10.7% absorption, showing that half of all the samples give absorption figures in this very small range of only 1.5%. This **inter-quartile range** is thus one possible measure of the dispersion or spread of the distribution.

Dividing the distribution at each successive 10% of the cumulative frequency yields sectional figures known as **deciles,** and similar division of a large distribution into 100 sections gives **percentiles.**

Calculation of the mean

Using the same set of figures for percentage absorption of 100 concrete blocks, the calculation would normally be set out in vertical columns as follows:

absorption %	central value x	frequency f	product xf
up to 8	7.5	1	7.5
up to 9	8.5	2	17.0
up to 10	9.5	8	76.0
up to 11	10.5	22	231.0
up to 12	11.5	35	402.5
up to 13	12.5	21	262.5
up to 14	13.5	7	94.5
up to 15	14.5	3	43.5
up to 16	15.5	1	15.5
	Mean = 11.5	100	1150.0

EXERCISE 13(a)

1. The following table gives the marks awarded to 100 students in an examination:

x	1–10	11–20	21–30	31–40	41–50	51–60	61–70	71–80	81–80	91–100
f	1	4	8	17	24	21	14	6	3	2

Draw the histogram of this distribution.

2. The following table shows the number of children in each of 500 families. Draw the histogram and the distribution curve for this set of figures. The distribution curve is not symmetrical but is skewed positively; show that the mode is less than the mean.

No. of children	0	1	2	3	4	5	6	7	8	9	10	more
Families	62	95	120	87	61	39	20	9	3	2	1	1

3. In the same examination, 22 students in one class gained an average of 46 marks, whilst 28 students in another class had an average of 51 marks. Calculate the overall average for the two classes together.

4. Calculate the ten-yearly moving averages for the following set of figures of the number of births (in thousands) in the United Kingdom. Plot the graph of the moving average and note that it is not a straight line but has a prominent 'bulge'.

Year	1933	1934	1935	1936	1937	1938	1939	1940
Births	692	712	711	720	724	736	727	702
Year	1941	1942	1943	1944	1945	1946	1947	1948
Births	696	772	811	878	796	955	1025	905
Year	1949	1950	1951	1952	1953	1954	1955	1956
Births	855	818	797	793	804	795	789	825

5. In the routine testing of samples on a large contract, the following figures were collected for crushing loads applied to 120 concrete cubes. The values are in newtons and are grouped to the nearest 500 N.

Load	6500	7000	7500	8000	8500	9000	9500	10 000
Frequency	3	9	21	24	27	18	12	6

Draw a cumulative frequency curve for this distribution and mark on it the 25th, 50th and 75th percentiles. Estimate the semi-interquartile range.

Standard deviation

The two sets of numbers 3, 4, 5, 6, 7 and 1, 3, 5, 7, 9 each have the same mean but it is obvious that the **spread** is different in the two cases. In this simple comparison it would be sufficient to use the **range** as a measure of dispersion, 3 to 7 and 1 to 9, but in general the range is not a very satisfactory guide as it is entirely dependent upon the two extreme items. The measure of dispersion which takes account of all the items in the distribution and is used as the basis for further statistical calculations is the **standard deviation.**

Basically, the standard deviation is the root-mean-square value of deviations from the mean. To calculate it, first find by how much each value of the variate, x, differs from the mean, then square these deviations from the mean, then calculate the mean of the squares of deviations, and, finally, take the square root.

The symbol σ is used for standard deviation, and its mathematical determination may be expressed as

$$\sigma = \sqrt{\frac{1}{n}\sum(x - \bar{x})}$$

where n is the total number of items and $\bar{x}$ is the arithmetic mean.

For a grouped distribution the standard deviation is given by:

$$\sigma = \sqrt{\frac{1}{n}\sum f(x - \bar{x})}$$

The square of the standard deviation (σ^2) is also used as a measure of the spread of a distribution and is known as the **variance.**

For comparing two sets of samples it is convenient to express the standard deviation as a percentage of the mean to give the **coefficient of variation.**

For the two sets of numbers 3, 4, 5, 6, 7 and 1, 3, 5, 7, 9 the calculation of the standard deviations may be set out conveniently as follows:

SET I			SET II		
Variate x	Deviation d	(Deviation)^2 d^2	Variate x	Deviation d	(Deviation)^2 d^2
3 (5 items)	−2	4	1	−4	16
4	−1	1	3	−2	4
5	0	0	5	0	0
6	1	1	7	2	4
7	2	4	9	4	16
5 | 25		5 | 10	5 | 25		5 | 40
$\bar{x} = 5$		2	$\bar{x} = 5$		8
	$\sigma = \sqrt{2}$				$\sigma = \sqrt{8} = 2\sqrt{2}$

Thus the standard deviation for the second set is double that for the first.

It is advisable to adopt a suitable tabular form for setting out the calculation of the mean and the standard deviation as illustrated by the following examples.

Example 1 The labour expenditure (man-hours per house) of twenty skilled bricklayers is given below. Calculate the mean and the standard deviation.

603	625	559	703	549	658	618	589	638	597
512	645	544	611	565	583	678	574	667	482

No. of hours x	Deviation d		(Deviation)^2 d^2
603	+ 3		9
625	+ 25		625
559		− 41	1681
703	+ 103		10609
549		− 51	2601
658	+ 58		3364
618	+ 18		324
589		− 11	121
638	+ 38		1444
597		− 3	9
512		− 88	7744
645	+ 45		2025
544		− 56	3136
611	+ 11		121
565		− 35	1225
583		− 17	289
678	+ 78		6084
574		− 26	676
667	+ 67		4489
482		− 118	13924
20 | 12000	+ 446	− 446	20 | 60500
$\bar{x} =$ 600 hours	check		3025

$$\sigma = \sqrt{3025} = 55 \text{ hours}$$

Example 2 Tests on 100 specimen wood cubes gave the following results for densities in kilograms per cubic metre.

Density in kg/m^3	No. with this density
400–449	2
450–499	3
500–549	10
550–599	21
600–649	30
650–699	15
700–749	9
750–799	6
800–849	3
850–899	1

Calculate the mean and the standard deviation for this distribution.

Here we have a grouped distribution and it is convenient to regard the group 400–449 as though it were the single central value 425, and it is these central values that we shall in fact use for our variates. This may introduce a slight inaccuracy but this is unavoidable unless the 100 original values are known, and even then it is such a lengthy and tedious operation to deal separately with each one that it is very seldom worthwhile.

The central values 425, 475, 525, etc. are all multiples of 25 and it is a very convenient simplification to adopt a **class interval** of 25, dividing by this at the start and multiplying our results by it at the end.

Density in kg/m³	Central value	In units of 25 kg/m³	Frequency	Product	Deviation from mean	(Deviation)²	Product
group		x	f	xf	d	d^2	fd^2
400–449	425	17	2	34	−8.2	67.24	134.48
450–499	475	19	3	57	−6.2	38.44	115.32
500–549	525	21	10	210	−4.2	17.64	176.40
550–599	575	23	21	483	−2.2	4.84	101.64
600–649	625	25	30	750	−0.2	0.04	1.20
650–699	675	27	15	405	1.8	3.24	48.60
700–749	725	29	9	261	3.8	14.44	129.96
750–799	775	31	6	186	5.8	33.64	201.84
800–849	825	33	3	99	7.8	60.84	182.52
850–899	875	35	1	35	9.8	96.04	96.04
TOTALS	—	—	100	2520	—	—	1188.00

The complete calculation of mean and standard deviation is therefore as shown in the table; from the table

$$n = 100 \qquad \bar{x} = \frac{2520}{100} \times 25 = \underline{630 \text{ kg/m}^3}$$

$$\sigma = \sqrt{\frac{1188.00}{100}} \times 25 = \underline{86.2 \text{ kg/m}^3}$$

Assumed mean

It frequently happens that the value of the mean is an awkward decimal which makes the calculation of the standard deviation rather tedious. It is then better to assume that the mean does actually come to a convenient whole number, to work on this basis to calculate the standard deviation, and then to apply the appropriate correction afterwards.

Example: Consider the previous example with an assumed mean of 625, i.e. with a value of 25 instead of the calculated value of 25.2 (with a class interval of 25 as before). The complete determination could then be set out in the following way:

Density in kg/m³	Central value	In units of 25 kg/m³	Frequency	Deviation from assmd. mean	Product	(Deviation)²	Product
			f	D	fD	D^2	fD^2
400–449	425	17	2	−8	−16	64	128
450–499	475	19	3	−6	−18	36	108
500–549	525	21	10	−4	−40	16	160
550–599	575	23	21	−2	−42	4	84
600–649	625	25	30	0	0	0	0
650–699	675	27	15	2	30	4	60
700–749	725	29	9	4	36	16	144
750–799	775	31	6	6	36	36	216
800–849	825	33	3	8	24	64	192
850–899	875	35	1	10	10	100	100
TOTALS	—	—	100	—	136—116	—	1192

$$n = 100 \qquad \bar{x} = \left(\frac{136 - 116}{100} + 25\right) \times 25 = \underline{630 \text{ kg/m}^3}$$

$$\sigma = \sqrt{\frac{\sum fD^2}{n} - \left(\frac{\sum fD}{n}\right)^2} \times \text{class interval}$$

$$= \sqrt{11.92 - (0.2)^2} \times 25$$

$$= \underline{86.2 \text{ kg/m}^3}$$

The normal distribution curve

Strictly speaking, very few distributions follow the 'normal' curve exactly, but unless the departure is pronounced it is useful to assume that many of the continuous distributions follow it approximately. Other types of distributions such as the binomial, Poisson, bimodal, and skew distributions are dealt with in text books on statistics and are beyond the scope of this present volume.

For a normal distribution curve there should be complete symmetry and it follows that the mean, the mode and the median should all coincide. For purposes of comparison it is convenient mathematically to take the value of the mean as the origin of co-ordinates and measure along the horizontal axis in intervals of σ as shown in fig. 59.

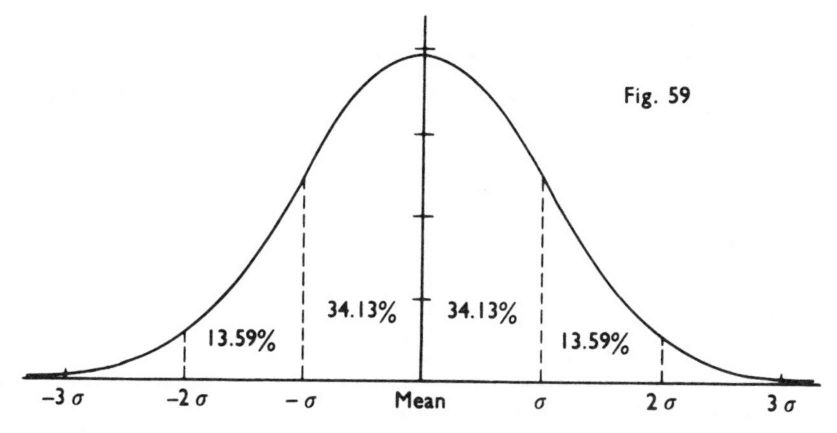

Although theoretically the normal curve extends indefinitely each way, it can be shown that:

50	per cent of the items come between the limits				$\pm$ 0.6745σ
68.26	,,	,,	,,	,,	$\pm$ σ
86.64	,,	,,	,,	,,	$\pm$ 1.5σ
95	,,	,,	,,	,,	$\pm$ 1.96σ
95.44	,,	,,	,,	,,	$\pm$ 2σ

The equation of the normal curve can be expressed in various forms, but the exponential functions involved are beyond the scope of this book. However, noting the following facts is helpful when sketching a normal curve (fig. 60):

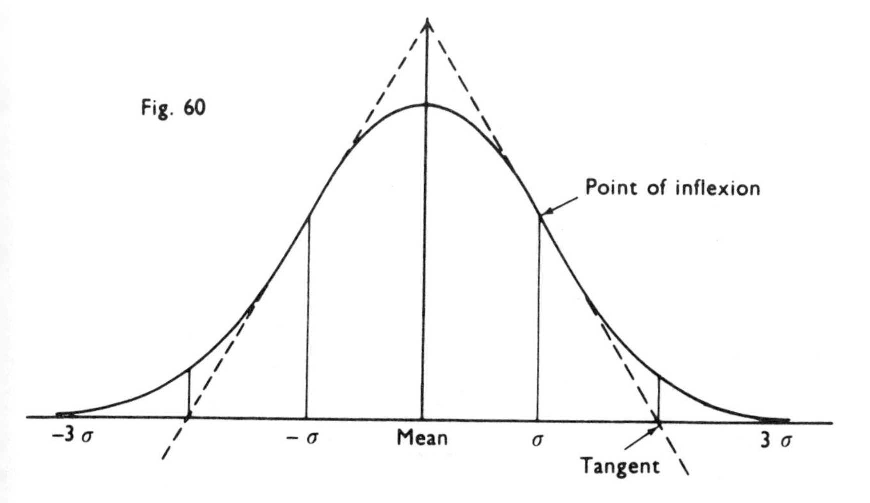

(i) The curve is symmetrical about the mean, and is often shown with the y-axis at this central value.

(ii) The curve never meets the x-axis at an angle, but approaches closer and closer to the x-axis at either end.

(iii) The places on the curve where the curvature changes (points of inflexion) occur where x has the values $\pm\sigma$.

(iv) The slope of the curve at these points is such that the tangent to the curve where $x = \sigma$ should cross the x-axis where $x = 2\sigma$, and the corresponding tangent on the other side where $x = -2\sigma$.

Area under the normal curve

In the form of the normal *frequency* distribution curve, the area between the curve and the x-axis is equal to the total number of items, observations, or results.

For probability estimations an alternative form is used for the normal *probability* distribution curve in which the total area under the curve is equal to unity (fig. 61). Table A on page 178 gives figures for the area under this curve between the mean and an ordinate at z such that

$$z = \frac{x - \bar{x}}{\sigma}$$

This form of the table is in common use for estimating probabilities and the examples which are given should be followed carefully until the method is grasped. Table B (p. 180) is an alternative form of this and gives the percentage of the area of the curve lying beyond any particular ordinate.

Normal probability distribution curve

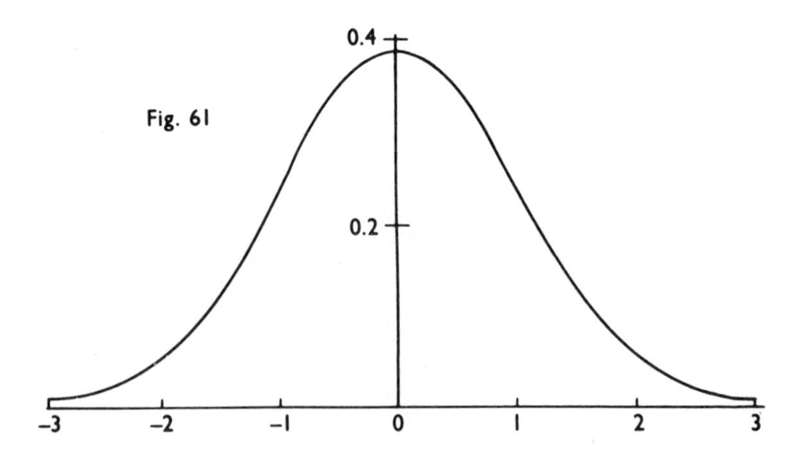

Example 1 Find the area under the normal curve between $z = 0$ and $z = 0.70$.

In Table A, follow down the first column until you reach the value $z = 0.7$. The figure for 0.70 is 0.2580 and this is the required area. It represents the probability that z lies between 0 and 0.70.

Example 2 Find the probability of a value of z between 0.95 and 3.05.

From Table A,

0 to 3.05 has probability	0.4989
0 to 0.95 has probability	0.3289
subtracting	0.1700

$\therefore$ 0.95 to 3.05 has probability 0.17

Example 3 Find the probability of a value of $z < -0.51$.

From Table A,

0 to 0.51 has probability	0.1950
-0.51 to 0 has probability	0.1950 because of symmetry
area below -0.51 is	$0.5000 - 0.1950 = 0.3050$

and the required probability is 0.305

Example 4 Find the area under the normal curve between the limits of ± 0.44.

From Table A,

the required area is double that for $z = 0.44$

i.e.

$$2 \times 0.1700 = 0.34$$

Example 5 If 1000 students sit an examination and the results are standardised to a mean of 50% and a standard deviation of 10%, estimate the number of students gaining over 75% and the number less than 40%.

$75\% = 50\% + 2.5 \times 10\% = \bar{x} + 2.5\sigma$

0 to 2.5 includes 0.4938

$\therefore$ above 75% is $0.5000 - 0.4938 = 0.0062$

and no. of students is $1000 \times 0.0062 =$ 6 students

$40\% = 50\% - 10\% = \bar{x} - \sigma$

0 to -1.0 includes 0.3413, so below 40% is $0.5000 - 0.3413 = 0.1587$

and no. of students is 1000×0.1587, i.e. about 159 students.

Note: Table B could also be used for this question and yields the same results.

For 75% mark, look up 2.5σ, giving 0.621%

and 0.621% of 1000 = 6 students

For 40% mark, look up 1.0σ, giving 15.87%

and 15.87% of 1000 = 159 students.

Example 6 Find the percentage of the area under the normal curve between the limits of 3σ either side of the mean.

For 3σ, Table B shows 0.135% above this limit;
for -3σ, similarly, 0.135% lie below;
i.e. 0.27% excluded and 99.73 included.

Note that this example used Table B, and the symmetry of the normal curve which implies that of the percentage lying outside the stated limits, half will lie above and half below. Thus for the timber specimens for which the mean was 630 kg/m³ and the standard deviation 86.2 kg/m³, $2\frac{1}{2}$%, i.e. 1 in 40, will have a density less than $630 - 1.96 \times 86.2 = 461$ kg/m³, but only 1 in 1000 will have a density of less than 370 kg/m³. Note that this does not mean that if exactly 1000 wood cubes were tested, that one and only one of these would have a density of less than 370 kg/m³, but it does mean that if many thousands of such specimens were tested, an average of approximately one per thousand could almost certainly be expected to be so light.

We are only justified in using Table A or Table B when we have a distribution which is approximately 'normal'. To determine whether a particular set of observations can be so regarded, it is useful to use probability paper and plot the points we would use for a cumulative frequency curve (ogive) on

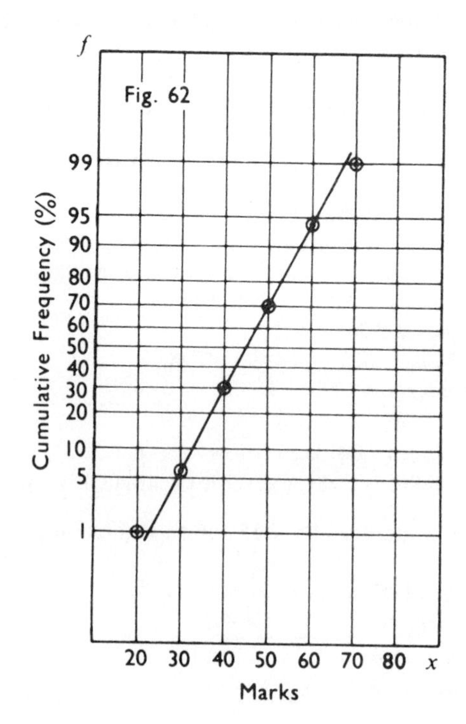

this special graph paper which has the effect of straightening the familiar S-shaped curve (see fig. 58).

If the result approximates to a straight line graph, then the distribution can be regarded as sufficiently close to a normal distribution to enable us to use the standard tables.

Example The results of an examination taken by a large number of students were analysed and the percentage of the students who gained marks up to various levels were recorded as shown in the following table. Plot these figures on probability paper to determine whether the distribution can be regarded as normal.

x	Marks gained	up to 20	to 30	to 40	to 50	to 60	to 70	up to 100
f	% of students	1	6	30	70	94	99	100

Strength of materials

As the central group of points lie close to a straight line when plotted on probability graph paper (fig. 62), the frequency distribution approximates to a normal distribution.

At the start of this chapter we discussed the case of forty wood specimens with strengths from 28 to 56 N/mm^2 and a mean of 42 N/mm^2. Knowing the full details of the tests, we could calculate the standard deviation, but for simplicity let us suppose that σ = 5 N/mm^2. Using this figure we can now refer to Table A to estimate the probability of finding a timber sample with a strength below 28 N/mm^2 (assuming that our strength figures are normally distributed).

Since

$$\bar{x} = 42 \quad \text{and} \quad \sigma = 5, \qquad 28 = \bar{x} - 2.8\sigma$$

Probability of 0 to 2.8 is 0.4974 (Table A)

Probability of a value below -2.8σ is 0.5000 − 0.4974

i.e. 0.0026

This indicates that for this particular softwood, specimens with a strength less than 28 N/mm^2 might be expected to occur at the rate of approximately 1 in 385.

Checking for another sample with the same range and the same mean value but with a standard deviation of 4 N/mm^2 yields a probability of 0.000 23 and a corresponding frequency of less than 1 in 4000.

If laminated construction is employed, it is extremely unlikely that each layer of the laminate will have a strength in the very low part of the complete range of possible values and thus, in general, a properly bonded laminated construction should be more reliable than its equivalent in solid timber. Statistically this means that although the mean is unaltered by the lamination, the standard deviation is reduced, and thus the risk of failure at a given stress is also reduced. Approximately, if m is the number of members in the laminate, the standard deviation of the laminated construction is given by:

$$\sigma_m = \frac{\sigma}{\sqrt{m}}$$

Quality control

Statistical methods are used in quality control to analyse the effect on the quality of a particular product (such as concrete) of any changes made in the course of production. The analysis may be done afterwards when all the results are available and the methods of this chapter will yield information which may be useful in respect of subsequent contracts. The analysis may also be attempted whilst the work is still in progress to try to discover whether the quality of the product is being maintained within required limits.

The commonest application is in the testing of concrete cubes after a set time such as 7 days or 28 days. The quality is defined in terms of a minimum compressive strength below which only a certain specified small proportion of results may fall. Continuous testing should reveal whether the average strength of the concrete being produced is below the required standard and this will indicate whether the proportions used in the mix need to be modified. Also, calculating the standard deviation when enough results are available should show if the quality is too variable and tighter control necessary. Analysing the results in short sequences of, say, groups of 10, should indicate any trend, and, if there is a tendency for the strength of the concrete to be deteriorating, this can be corrected. Similarly, the same methods can be used to prevent the wasteful use of expensive materials if it is shown that the mix is richer than required to satisfy the specifications.

Example A large number of concrete cubes were tested and the overall result was a mean strength of 32.5 N/mm² and a standard deviation of 2.8 N/mm².

Find the value below which less than 1 in 100 should fall.

Table B gives approx. 2.33σ for the 1% level

Hence, required value is $32.5 - 2.8 \times 2.33 = \underline{26.0}$ N/mm²

Units of stress

For expressing force over an area, the appropriate SI unit should be N/m². When measuring high stress values as in concrete cube tests, multiples of this basic unit are involved and these stresses of millions of newtons per metre squared could be expressed in MN/m², i.e. mega-newtons per metre squared.

This multiple unit was not favoured by the British Standards Institution who suggested the use of the equivalent form of N/mm², i.e. newtons per millimetre squared. Recently, a new unit has been proposed called the pascal for which the symbol is Pa. The pascal is another name for the newton per metre squared, i.e. 1 Pa = 1 N/m^2. It follows that stress units may now be expressed in pascals, kilopascals, or megapascals. The units of stress for values derived from concrete cube tests may therefore be expressed in megapascals and it is likely that this unit will gradually gain acceptance. Meanwhile it should be noted that

$$1 \text{ MPa} = 1 \text{ MN/m}^2 = 1 \text{ N/mm}^2$$

and examination questions may be set with, or may have used, any of these three forms.

Symbols

To avoid confusion, no attempt has been made to complicate the basic concepts of statistics introduced in this chapter. I have been reluctant to include any material on sampling or confidence limits or other subjects outside the ONC Construction Syllabus. However, for the reader who may now wish to proceed to further studies in statistics, I should point out that when it is necessary to distinguish between the mean of a sample and the mean of a total population, the sample mean is denoted by $\bar{x}$ and the population mean by μ. Similarly, the standard deviation of the sample is denoted by s and σ implies the standard deviation of the population from which the sample is extracted.

EXERCISE 13(b)

1. Extensions to an existing public building are to be carried out in brickwork matching the original as near as possible. Sample measurements are taken of the combined length of three bricks and three joints (in mm) as follows:

685	715	695	665	725	705	675	695	705	715
725	695	665	705	695	655	665	675	685	655

Construct a frequency distribution table for these results, and find their mean and standard deviation.

2. In absorption tests on 200 bricks the following figures were obtained:

% absorption	6	7	8	9	10	11	12	13	14	15
frequency	1	5	13	31	51	47	33	14	4	1

Calculate the mean and the standard deviation for these results.

3. The number of hours worked in a certain week by each of 100 employees is recorded as follows:

Hours per man	40	41	42	43	44	45	46	47	48
No. of men	3	4	8	12	17	21	19	11	5

Construct a histogram for these results. Calculate the values of the mean, the median and the standard deviation.

4. The table below shows the wages paid by a firm to 1000 of its employees:

Wages £	*No. of* employees
22–24	53
24–26	73
26–28	140
28–30	190
30–32	209
32–34	170
34–36	86
36–38	57
38–40	22

Calculate the mean wage and the standard deviation from it.

5. The following table gives the breaking stress of each of 10 test cubes of concrete:

Cube	1	2	3	4	5	6	7	8	9	10
Breaking stress N/mm²	19.0	25.2	21.7	23.8	25.9	20.3	21.0	26.6	22.4	23.1

Calculate the mean breaking stress, and the standard deviation of the given set of values. What do you understand by this number which you have derived for the standard deviation? (D.D.C.T.)

6. The following table gives the number of pieces of timber having a particular length in a sample of 100 pieces of timber.

Length of piece of timber (mm)	1100	1120	1140	1160	1180	1200	1200
Number having this length	2	4	12	28	38	10	6

Calculate the mean and standard deviation of the given sample and state what you mean by the term standard deviation. (D.D.C.T.)

7. If it is known that in an examination 33% of the candidates scored at least 60 marks but only 9% scored 78 marks or more, estimate the mean and the standard deviation on the assumption that the distribution of scores is normal.

8. Under a certain normal curve, 33% of the area represents values of $x \leqslant 50$ and 2.5% represents $x \geqslant 74$. Find the mean and standard deviation of this distribution.

9. The mean life of a masonry drill bit is 20 hours active use. Assuming a normal distribution and a standard deviation of 3 hours, what is the probability of a particular bit lasting longer than 24 hours? If another drill bit which was supposed to be identical lasted only 10 hours, would a complaint to the manufacturer be justified?

10. The following table shows the results of crushing tests on a number of concrete cubes tested in the laboratory.

Draw a histogram for these results and read off the Mode.

Calculate the value of the mean and standard deviation.

Crushing stress (MPa)	20	22	24	26	30	32	34
No. of cubes failing at this stress	2	7	20	19	4	3	1

REVISION EXERCISE 13

1. The table below shows (in thousands) the number of houses completed in this country over a period of 20 consecutive months:

Months	1	2	3	4	5	6	7	8	9	10
Houses	13.5	14.6	19.0	15.4	17.3	17.0	18.2	15.6	19.7	20.5
Months	11	12	13	14	15	16	17	18	19	20
Houses	19.5	19.7	17.8	17.5	25.0	20.2	22.8	22.5	24.9	24.7

Plot these figures on a graph and superimpose the graph of the moving average.

2. One hundred trainees were tested on a job and their times to complete it were recorded as follows:

Time in minutes	7–7½	7½–8	8–8½	8½–9	9–9½	9½–10
No. of trainees	13	17	28	20	15	7

Illustrate these statistics by means of a histogram and determine the mean time taken, the median and the mode. (E.M.E.U.)

3. In checking the thicknesses of a sample of 100 slates, the following details were noted:

Thickness in mm	5.0	5.2	5.4	5.6	5.8	6.0	6.2	6.4	6.6	6.8	7.0
Number of slates	1	1	1	5	20	31	25	10	3	2	1

Find the mean, median, mode and standard deviation from the mean of this distribution. (E.M.E.U.)

4. (a) Twenty observations of an angle, made under similar conditions, are listed below. Determine the mean observation.

Observation	67° 30′ 10″	67° 30′ 12″	67° 30′ 13″	67° 30′ 14″
Frequency	1	1	2	4
Observation	67° 30′ 15″	67° 30′ 16″	67° 30′ 17″	67° 30′ 18″
Frequency	6	2	2	2

(b) By adding 6 to each of the numbers 3, 5, 6, 2, 1, 7, the set 9, 11, 12, 8, 7, 13 is obtained. Show that these two sets have the same standard deviation but different means and evaluate the two means and the common standard deviation. (U.L.C.I.)

5. The following table gives the percentage moisture content in 100 test samples of a particular cement mix, after a certain time interval.

Per cent moisture content	22	24	26	28	30	32	34
No. with this moisture content	2	4	13	17	52	11	1

Calculate the mean moisture content, and the standard deviation. State what you understand by the term standard deviation. (D.D.C.T.)

6. In a 10 week period, the number of men employed per week on a small construction site was 2, 2, 3, 3, 3, 4, 5, 6, 6, 7.
   Use this set of figures to calculate:
   (i) the median,
   (ii) the mode,
   (iii) the mean,
   (iv) the standard deviation. (U.E.I.)

7. During a contract the following table was compiled from results of tests on works concrete cubes:

Crushing strength in N/mm²	14.0	14.1	14.2	14.3	14.4	14.5	14.6	14.7	14.8
No. of cubes falling	1	5	8	15	25	12	7	5	2

   (i) Construct a histogram for these results.
   (ii) Determine the median, mode and mean strength. (N.C.T.E.C.)

8. The average gross pay, correct to the nearest £1, of a group of workers was recorded for 13 consecutive weeks, and the results were as follows:

   £23, 21, 23, 18, 22, 25, 27, 29, 29, 25, 22, 26, 22.

   Calculate, for this set of figures,
   (i) the mean,
   (ii) the median,
   (iii) the mode,
   (iv) the standard deviation. (U.E.I.)

9. The following table shows the number of hours worked by each of 100 employees in a particular week.

Hours per man	38	39	40	41	42	43	44	45	46
No. of men	3	4	8	12	17	21	19	11	5

   Calculate the values of mean, median and standard deviation. (U.E.I.)

10. 100 apprentices are each given a similar practical test to complete and all 100 start at the same time. Every two minutes the number of completed tests is counted, and the figures are as follows:

Time in minutes	2	4	6	8	10	12	14	16	18	20
No. completed	0	0	0	0	4	19	62	92	98	100

   Find the mean time taken and the standard deviation from the mean.
   Estimate the inter-quartile range. (U.E.I.)

11. (a) Calculate the angles in the triangle whose sides are 10 m, 8 m and 17 m.
    (b) The following readings were taken on to a staff placed at various points on supposedly level ground from a level at a fixed position:

    1.66, 1.65, 1.67, 1.67, 1.66, 1.64, 1.68, 1.67, 1.66, 1.65 (all metres),

    Calculate the standard deviation of these readings. (N.C.T.E.C.)

12. The following table gives ordinates of the standard normal curve.
The distribution it represents has a mean of zero and unit standard deviation.

x	0	± 0.5	± 1.0	± 1.5	± 2.0	± 2.5	± 3.0	± 3.5	± 4.0
y	.399	.352	.242	.130	.054	.017	.004	.001	0

(i) Plot the curve for values of x from -4 to $+4$.
(ii) Use Simpson's Rule to calculate the area under the curve for values of x between the limits of ± 1.5.
(iii) For an examination marked out of 100, it is found that the results give a mean of 60 and a standard deviation of 10. Find the percentage of students who failed to reach the pass mark of 45. (U.E.I.)

13. Plot the distribution curve given by the following set of values

Score value x	0	2	4	6	8	10	12	14	16	18	20
Frequency f	0	0	4	55	244	394	244	55	4	0	0

(a) Calculate the mean and standard deviation.
(b) Find the area under the curve for values of x from 6 to 14.
(*Either tables of area under the normal curve or Simpson's Rule may be used.*)
(U.E.I.)

Statistical Tables

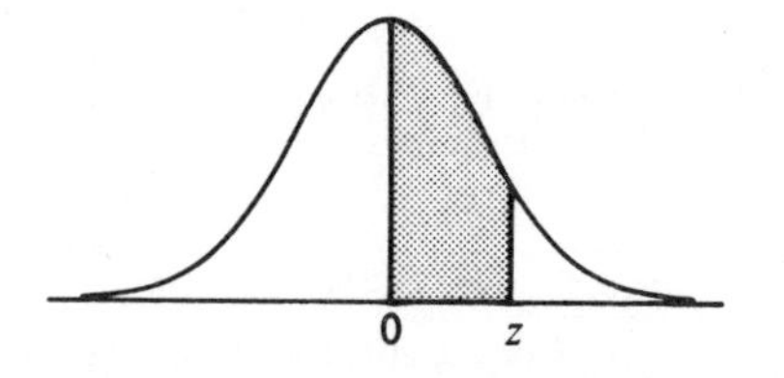

TABLE A. AREAS UNDER THE NORMAL CURVE
(To three decimal places)

z	0.00	0.01	0.02	0.03	0.04	0.05	0.06	0.07	0.08	0.09
0.0	0.000	0.004	0.008	0.012	0.016	0.020	0.024	0.028	0.032	0.036
0.1	0.040	0.044	0.048	0.052	0.056	0.060	0.064	0.067	0.071	0.075
0.2	0.079	0.083	0.087	0.091	0.095	0.099	0.103	0.106	0.110	0.114
0.3	0.118	0.122	0.126	0.129	0.133	0.137	0.141	0.144	0.148	0.152
0.4	0.155	0.159	0.163	0.166	0.170	0.174	0.177	0.181	0.184	0.188
0.5	0.191	0.195	0.198	0.202	0.205	0.209	0.212	0.216	0.219	0.222
0.6	0.226	0.229	0.232	0.236	0.239	0.242	0.245	0.249	0.252	0.255
0.7	0.258	0.261	0.264	0.267	0.270	0.273	0.276	0.279	0.282	0.285
0.8	0.288	0.291	0.294	0.297	0.300	0.302	0.305	0.308	0.311	0.313
0.9	0.316	0.319	0.321	0.324	0.326	0.329	0.331	0.334	0.336	0.339
1.0	0.341	0.344	0.346	0.348	0.351	0.353	0.355	0.358	0.360	0.362
1.1	0.364	0.367	0.369	0.371	0.373	0.375	0.377	0.379	0.381	0.383
1.2	0.385	0.387	0.389	0.391	0.393	0.394	0.396	0.398	0.400	0.401
1.3	0.403	0.405	0.407	0.408	0.410	0.411	0.413	0.415	0.416	0.418
1.4	0.419	0.421	0.422	0.424	0.425	0.426	0.428	0.429	0.431	0.432
1.5	0.433	0.434	0.436	0.437	0.438	0.439	0.441	0.442	0.443	0.444
1.6	0.445	0.446	0.447	0.448	0.449	0.451	0.452	0.453	0.454	0.454
1.7	0.455	0.456	0.457	0.458	0.459	0.460	0.461	0.462	0.462	0.463
1.8	0.464	0.465	0.466	0.466	0.467	0.468	0.469	0.469	0.470	0.471
1.9	0.471	0.472	0.473	0.473	0.474	0.474	0.475	0.476	0.476	0.477
2.0	0.477	0.478	0.478	0.479	0.479	0.480	0.480	0.481	0.481	0.482
2.1	0.482	0.483	0.483	0.483	0.484	0.484	0.485	0.485	0.485	0.486
2.2	0.486	0.486	0.487	0.487	0.487	0.488	0.488	0.488	0.489	0.489
2.3	0.489	0.490	0.490	0.490	0.490	0.491	0.491	0.491	0.491	0.492
2.4	0.492	0.492	0.492	0.493	0.492	0.493	0.493	0.493	0.493	0.494
2.5	0.494	0.494	0.494	0.494	0.494	0.495	0.495	0.495	0.495	0.495
2.6	0.495	0.495	0.496	0.496	0.496	0.496	0.496	0.496	0.496	0.496
2.7	0.497	0.497	0.497	0.497	0.497	0.497	0.497	0.497	0.497	0.497
2.8	0.497	0.498	0.498	0.498	0.498	0.498	0.498	0.498	0.498	0.498
2.9	0.498	0.498	0.498	0.498	0.498	0.498	0.499	0.499	0.499	0.499
3.0	0.499	0.499	0.499	0.499	0.499	0.499	0.499	0.499	0.499	0.499
3.1	0.499	0.499	0.499	0.499	0.499	0.499	0.499	0.499	0.499	0.499
3.2	0.499	0.499	0.499	0.499	0.499	0.499	0.499	0.499	0.499	0.499
3.3	0.500	0.500	0.500	0.500	0.500	0.500	0.500	0.500	0.500	0.500
3.4	0.500	0.500	0.500	0.500	0.500	0.500	0.500	0.500	0.500	0.500

TABLE A. AREAS UNDER THE NORMAL CURVE
(To greater accuracy)

Z	0.00	0.01	0.02	0.03	0.04	0.05	0.06	0.07	0.08	0.09
0.0	0.0000	0.0040	0.0080	0.0120	0.0160	0.0199	0.0239	0.0279	0.0319	0.0359
0.1	0.0398	0.0438	0.0478	0.0517	0.0557	0.0596	0.0636	0.0657	0.0714	0.0753
0.2	0.0793	0.0832	0.0871	0.0910	0.0948	0.0987	0.1026	0.1064	0.1103	0.1141
0.3	0.1179	0.1217	0.1255	0.1293	0.1331	0.1368	0.1406	0.1443	0.1480	0.1517
0.4	0.1554	0.1591	0.1628	0.1664	0.1700	0.1736	0.1772	0.1808	0.1844	0.1879
0.5	0.1915	0.1950	0.1985	0.2019	0.2054	0.2088	0.2123	0.2157	0.2190	0.2224
0.6	0.2257	0.2291	0.2324	0.2357	0.2389	0.2422	0.2454	0.2486	0.2517	0.2549
0.7	0.2580	0.2611	0.2642	0.2673	0.2704	0.2734	0.2764	0.2794	0.2823	0.2852
0.8	0.2881	0.2910	0.2939	0.2967	0.2995	0.3023	0.3051	0.3078	0.3106	0.3133
0.9	0.3159	0.3186	0.3212	0.3238	0.3264	0.3289	0.3315	0.3340	0.3365	0.3389
1.0	0.3413	0.3438	0.3461	0.3485	0.3508	0.3531	0.3554	0.3577	0.3599	0.3621
1.1	0.3643	0.3665	0.3686	0.3708	0.3729	0.3749	0.3770	0.3790	0.3810	0.3830
1.2	0.3849	0.3869	0.3888	0.3907	0.3925	0.3944	0.3962	0.3980	0.3997	0.4015
1.3	0.4032	0.4049	0.4066	0.4082	0.4099	0.4115	0.4131	0.4147	0.4162	0.4177
1.4	0.4192	0.4207	0.4222	0.4236	0.4251	0.4265	0.4279	0.4292	0.4306	0.4319
1.5	0.4332	0.4345	0.4357	0.4370	0.4382	0.4394	0.4406	0.4418	0.4429	0.4441
1.6	0.4452	0.4463	0.4474	0.4484	0.4495	0.4505	0.4515	0.4525	0.4535	0.4545
1.7	0.4554	0.4564	0.4573	0.4582	0.4591	0.4599	0.4608	0.4616	0.4625	0.4633
1.8	0.4641	0.4649	0.4656	0.4664	0.4671	0.4678	0.4686	0.4693	0.4699	0.4706
1.9	0.4713	0.4719	0.4726	0.4732	0.4738	0.4744	0.4750	0.4756	0.4761	0.4767
2.0	0.4772	0.4778	0.4783	0.4788	0.4793	0.4798	0.4803	0.4808	0.4812	0.4817
2.1	0.4821	0.4826	0.4830	0.4834	0.4838	0.4842	0.4846	0.4850	0.4854	0.4857
2.2	0.4861	0.4864	0.4868	0.4871	0.4875	0.4878	0.4881	0.4884	0.4887	0.4890
2.3	0.4893	0.4896	0.4898	0.4901	0.4904	0.4906	0.4909	0.4911	0.4913	0.4916
2.4	0.4918	0.4920	0.4922	0.4925	0.4927	0.4929	0.4931	0.4932	0.4934	0.4936
2.5	0.4938	0.4940	0.4941	0.4943	0.4945	0.4946	0.4948	0.4949	0.4951	0.4952
2.6	0.4953	0.4955	0.4956	0.4957	0.4959	0.4960	0.4961	0.4962	0.4963	0.4964
2.7	0.4965	0.4966	0.4967	0.4968	0.4969	0.4970	0.4971	0.4972	0.4973	0.4974
2.8	0.4974	0.4975	0.4976	0.4977	0.4977	0.4978	0.4979	0.4979	0.4980	0.4981
2.9	0.4981	0.4982	0.4983	0.4983	0.4984	0.4984	0.4985	0.4985	0.4986	0.4986
3.0	0.4986	0.4987	0.4987	0.4988	0.4988	0.4989	0.4989	0.4889	0.4990	0.4990
3.1	0.4990	0.4991	0.4991	0.4991	0.4992	0.4992	0.4992	0.4992	0.4993	0.4993
3.2	0.4993	0.4993	0.4994	0.4994	0.4994	0.4994	0.4994	0.4995	0.4995	0.4995
3.3	0.4995	0.4995	0.4996	0.4996	0.4996	0.4996	0.4996	0.4996	0.4996	0.4997
3.4	0.4997	0.4997	0.4997	0.4997	0.4997	0.4997	0.4997	0.4997	0.4997	0.4998
3.5	0.49977					0.49980				
3.6	0.49984					0.49987				
3.7	0.49990					0.49991				
3.8	0.49993					0.49994				
3.9	0.49995					0.49996				
4.0	0.49997									

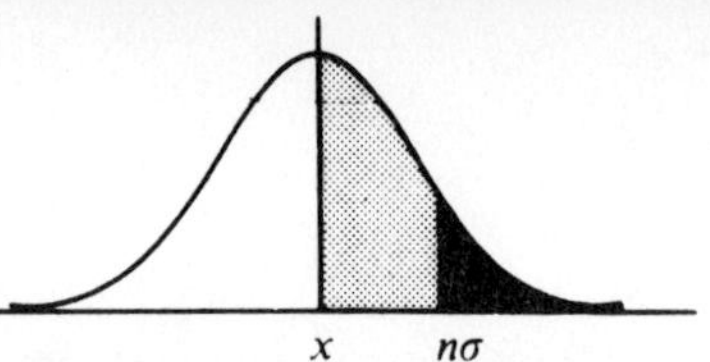

TABLE B. PERCENTAGE ABOVE A GIVEN VALUE OF $\bar{x} = n\sigma$

n	0.00	0.01	0.02	0.03	0.04	0.05	0.06	0.07	0.08	0.09
0	50.00	49.60	49.20	48.80	48.40	48.01	47.61	47.21	46.81	46.41
0.1	46.02	45.62	45.22	44.83	44.43	44.04	43.64	43.25	42.86	42.47
0.2	42.07	41.68	41.29	40.90	40.52	40.13	39.74	39.36	38.97	38.59
0.3	38.21	37.83	37.45	37.07	36.69	36.32	35.94	35.57	35.20	34.83
0.4	34.46	34.09	33.72	33.36	33.00	32.64	32.28	31.92	31.56	31.21
0.5	30.85	30.50	30.15	29.81	29.46	29.12	28.77	28.43	28.10	27.76
0.6	27.43	27.09	26.76	26.43	26.11	25.78	25.46	25.14	24.83	24.51
0.7	24.20	23.89	23.58	23.27	22.96	22.66	22.36	22.06	21.77	21.48
0.8	21.19	20.90	20.61	20.33	20.05	19.77	19.49	19.22	18.94	18.67
0.9	18.41	18.14	17.88	17.62	17.36	17.11	16.85	16.60	16.35	16.11
1.0	15.87	15.62	15.39	15.15	14.92	14.69	14.46	14.23	14.01	13.79
1.1	13.57	13.35	13.14	12.92	12.71	12.51	12.30	12.10	11.90	11.70
1.2	11.51	11.31	11.12	10.93	10.75	10.56	10.38	10.20	10.03	9.85
1.3	9.68	9.51	9.34	9.18	9.01	8.85	8.69	8.53	8.38	8.23
1.4	8.08	7.93	7.78	7.64	7.49	7.35	7.21	7.08	6.94	6.81
1.5	6.68	6.55	6.43	6.30	6.18	6.06	5.94	5.82	5.71	5.59
1.6	5.48	5.37	5.26	5.16	5.05	4.95	4.85	4.75	4.65	4.55
1.7	4.46	4.36	4.27	4.18	4.09	4.01	3.92	3.84	3.75	3.67
1.8	3.59	3.51	3.44	3.36	3.29	3.22	3.14	3.07	3.01	2.94
1.9	2.87	2.81	2.74	2.68	2.62	2.56	2.50	2.44	2.39	2.33
2.0	2.28	2.22	2.17	2.12	2.07	2.02	1.97	1.92	1.88	1.83
2.1	1.79	1.74	1.70	1.66	1.62	1.58	1.54	1.50	1.46	1.43
2.2	1.39	1.36	1.32	1.29	1.25	1.22	1.19	1.16	1.13	1.10
2.3	1.07	1.04	1.02	0.990	0.964	0.939	0.914	0.889	0.866	0.842
2.4	0.820	0.798	0.776	0.755	0.734	0.714	0.695	0.676	0.657	0.639
2.5	0.621	0.604	0.587	0.570	0.554	0.539	0.523	0.508	0.494	0.480
2.6	0.466	0.453	0.440	0.427	0.415	0.402	0.391	0.379	0.368	0.357
2.7	0.347	0.336	0.326	0.317	0.307	0.298	0.289	0.280	0.272	0.264
2.8	0.256	0.248	0.240	0.233	0.226	0.219	0.212	0.205	0.199	0.193
2.9	0.187	0.181	0.175	0.169	0.164	0.159	0.154	0.149	0.144	0.139
3.0	0.1350					0.1144				
3.1	0.0968					0.0816				
3.2	0.0687					0.0577				
3.3	0.0483					0.0404				
3.4	0.0337					0.0280				
3.5	0.0233					0.0193				
3.6	0.0159					0.0131				
3.7	0.0108					0.00884				
3.8	0.00723					0.00591				
3.9	0.00481					0.00391				
4.0	0.00317									
—										
4.5	0.00034									
—										
5.0	0.00003									

Answers to Exercises

Page 6 — EXERCISE 1(a)

1. 5	2. 4	3. 3	4. -2	5. 3
6. -1	7. 1	8. 4	9. -6	10. $\frac{1}{5}$
11. 2	12. -2	13. $-\frac{1}{3}$	14. 2	15. ± 4
16. ± 3	17. 0	18. ± 2	19. ± 2	20. 2

21. $r = \sqrt{\dfrac{3V}{\pi h}}$

22. $r = \dfrac{100I}{Pn}$

23. $h = \dfrac{S - 2\pi r^2}{2\pi r}$

24. $R = \sqrt{\dfrac{E^2}{I^2} - \omega^2 l^2}$

25. $d = \dfrac{2(S - an)}{n(n - 1)}$

26. $r = 6\sqrt{\dfrac{10A}{\pi\theta}}$

27. $D = \sqrt{\dfrac{6V}{\pi}}$

Page 11 — EXERCISE 1(b)

1.	(a) $\frac{3}{4}$	(b) $2y^2$	(c) 18	
2.	(a) $1\frac{1}{2}$	(b) $\frac{4}{5}$	(c) $\frac{4}{9}$	
3.	(a) 2^{-3}	(b) $2^{5/3}$	(c) $2^{4.32}$	(d) $2^{-0.678}$
4.	(a) 10^{-3}	(b) $10^{-2.523}$	(c) 10^{-2}	(d) $10^{-0.067}$
5.	(a) 4	(b) $1\frac{1}{3}$	(c) $-1\frac{1}{3}$	
6.	(a) 1.792	(b) 1.470	(c) -1.022	
7.	(a) $4\frac{1}{3}$	(b) $-1\frac{1}{3}$	(c) $\frac{1}{2}$	
8.	(a) 2	(b) 3.5	(c) 1	
9.	(a) 7.45	(b) 1.71		
10.	(a) 31.62	(b) 8.454		
11.	(a) 0.879	(b) 28.6	(c) 1.23	
12.	(a) 2	(b) 1		
13.	(a) $1\frac{1}{3}$	(b) $\frac{1}{3}$		
14.	(a) 1.19	(b) 0.171	(c) 1.84	

Page 15

EXERCISE 1(c)

1. 6
2. $\dfrac{D^2}{D^2 - d^2}$; $\dfrac{4}{3}$
3. (a) 22.62 (b) 67.55 (c) 46.67 (d) 2.475
4. (a) $2\sqrt{2} - 3$ (b) $-4 - \sqrt{15}$ (c) $-3 - \sqrt{5}$
7. 6.1% too small

Page 16

REVISION EXERCISE 1

1. 3 2. $\frac{4}{3}$ 3. 1
4. 2 5. 3 6. 8
7. (a) $a = 4$ (b) $a(5a - b + 6c)$
8. $x = 7, y = -2$
9. $x = 3, y = -2$
10. (a) $r = \dfrac{E}{C} - R$ (b) $l = \dfrac{2EIy}{Wx^2} + \dfrac{x}{3}$
11. $t = D\left[\dfrac{P(m^2 - 1)}{2Em^2}\right]^{1/3}$
12. $D = \dfrac{ab^3SY}{3Wl^2} - \dfrac{L}{2}$
13. $K = k\left(\dfrac{Av}{aTu}\right)^2$
14. 0.682
15. (a) 14.5 (b) 1.30%
16. (a) 817 (b) 70 (c) 975
17. (a) 4.44×10^4 (b) 408
18. (a) $d = \dfrac{bn^2E_c}{2AE} + n$ (b) 7.32 or -11.55
19. 0.438
20. (a) $\dfrac{z(W - nW_2)}{ndW_2}$ (b) 11 520
21. 5.32
22. (a) $x = \sqrt{A - \dfrac{P^2}{Q^2}}$ (b) (i) 1.32 (ii) -0.268
23. (a) 2.24 (b) 912
24. (a) 0.583 (b) 5.85
25. 6.76
26. $P_2 = \dfrac{P_1V_1^n}{V_2^n}$; 0.89

27. (a) 0.523 (b) 15.3

28. $\frac{9}{4}a^{10}b^2$

29. (a) 0.0395 (b) $L = 2bdYt^2 - \frac{3}{8}Mg$

30. $p = \left(\frac{D^2 - d^2}{D^2 + d^2}\right) f$; 2.40

31. (a) 5.53 (b) 61.4 (c) 11.6

32. (a) $\frac{1}{2\pi R}\sqrt{S(4\pi R^2 - 5)}$ (b) 3.426 (c) $\frac{3}{10}$ (d) $\frac{4y}{3x}$

Page 24

EXERCISE 2(a)

1. (a) 41.7 m/s (b) 140.4 km/h
2. (a) $36\frac{1}{2}$p (b) 1312 g
3. (a) 2; $1\frac{1}{3}$ (b) -5; 7 (c) $1\frac{1}{3}$; $2\frac{1}{4}$
   (d) $\frac{1}{4}$; $-\frac{5}{8}$ (e) 3; $\frac{3}{8}$ (f) -1; 7
   (g) 2; 13 (h) $1\frac{1}{3}$; 4
4. $6\frac{8}{13}$ units²

Page 28

EXERCISE 2(b)

1. $a = 5, b = 1$
2. $m = \frac{1}{3}, c = 3, x = 4$
3. $a = 3, b = 2$
4. $d = \frac{2P}{3}$
5. £19
6. 18 years
7. 95.5
8. $a = 1, n = 4$
9. $a = -10, b = 320$

Page 30

REVISION EXERCISE 2

1. $a = \frac{1}{3}, b = -3$
2. $a = 3, b = 8, c = -15$
3. $A = 2\frac{1}{3}, B = -5\frac{1}{3}, C = 3$
4. $a = 0.2, b = 0.35$; $L = 4\frac{1}{3}, y = 3.35$
5. $a = 0.15, b = 2.5$
6. $a = 0.452, b = 1.85$
7. $m = 0.2, c = 25$; $w = 32$ g
8. $S = 35 - 4Q$
9. $1000d = 7w + 100$
10. 371
11. $A = 12.6, b = 0.4$
12. $a = 6.3, n = 0.5$
13. $V = 4.3$
14. $k = 1.6, n = 0.48$
15. $S = \frac{14\,000}{9^x}$
16. $m = 2.3, c = -0.78$

Page 40

EXERCISE 3(a)

1. (a) $(x + 2y)(2 - x)$ (b) $(3a + 2c)(2b - a)$
2. (a) $\dfrac{x^2}{y}$ (b) $\dfrac{4}{x^2 - 4}$
3. (a) $(3 + 2x)(3 - 2x)$ (b) $(\frac{1}{2}a + b^2)(\frac{1}{2}a - b^2)$
4. (a) $(x - 2)(x^2 + 2x + 4)$ (b) $(3 + xy^2)(9 - 3xy^2 + x^2y^4)$
5. (a) 3, 5 (b) 7, -4
   (c) $\frac{1}{3}$, 1 (d) $1\frac{1}{2}$, -4
   (e) 1, $1\frac{2}{3}$ (f) 0, $\frac{1}{4}$
   (g) 0, $\frac{9}{16}$ (h) 1, $-\frac{5}{8}$
   (i) $-\frac{1}{2}$ (j) $\frac{2}{3}$
6. (a) $\frac{3}{4}$, -2 (b) 0.90, -2.23
7. (a) -8.16, -1.84 (b) 1.70, 5.30
   (c) 1.09, -0.52 (d) 0.45, -1.12
   (e) 0.81, -0.61 (f) 2.69, -0.19
8. (a) $\frac{1}{3}$, -2 (b) 0, $\frac{1}{3}$, $\frac{1}{2}$
   (c) ± 1, ± 2 (d) 1, 4
   (e) 2, $-1\frac{1}{3}$ (f) -8, -15
9. (a) 0.32, 6.28 (b) 2.85, -0.35
   (c) 5, -2 (d) $\frac{2}{3}$, -1
   (e) $\frac{1}{4}$, $\frac{1}{2}$ (f) 0.46, 6.54
10. 9.74, 0.92
11. 2.82, 0.18; 3.41, 0.59
12. (a) 1.62, -0.62 (b) 0.73, -2.73
13. 5.83, 0.17
14. 4.45, -0.45; 3.73, 0.27; -4
15. (a) 1.30, -2.30 (b) 3.56, -0.56
16. (a) 4.83, -0.83 (b) 5.46, -1.46

Page 50

EXERCISE 3(b)

1. (a) $x = 2, y = 1\frac{1}{2}$
2. 375
3. $a = 5.8, b = -0.9$; $a = -13.8, b = 8.9$
4. $x = \frac{1}{2}, y = -1\frac{1}{2}$; $x = 1\frac{2}{3}, y = 2$
5. $x = +\sqrt{3}, -\sqrt{3}, +2, -2$; $y = 0, 0, -1, +1$
6. (a) $x = 2\frac{1}{3}, y = \frac{2}{3}$;
   $x = 1, y = -2$
   (b) $a = 2, b = -3, c = -9$; -10
7. $x = 6, y = 2$; $x = -2, y = 6$
8. $x = 1$, 3 or -2
9. $x = 2.67$, -2.15 or -0.52
10. $x = 1$, 4 or -3
11. $x = -3.63$
12. 8.20 m/s

Page 51 **REVISION EXERCISE 3**

1. $-1.27, -4.73$
2. $6.53, -1.53$
3. (a) $+1, -1;\ 1\frac{1}{3}, -2\frac{1}{2}$ (b) $x = -p \pm \sqrt{p^2 + q}$
4. $1.6, -0.3$
5. (a) $27, -2$; (b) $1\frac{2}{3}, -1\frac{1}{2}$
6. $7.33, -8.43$
7. (a) $14, -14$; (b) $11, 17$
8. $1.30, -2.30$
9. (a) $1, -2\frac{1}{3}$ (b) $2.43, -1.10$
10. $0.39, -3.89$
11. (a) $-\frac{1}{6}$ (b) $(1 - x)(1 - x + y)$; $(a + b)(a - b)(x + y)(x - y)$
12. $4, 12$
13. (a) $n = \dfrac{-am \pm \sqrt{am(am + 2bc)}}{b}$ (b) $14.6, -5.2$
14. (a) $2.45, -0.61$ (b) $2x + 3, 2x - 3$
15. (a) $1.55, -0.22$ (b) $x = 1$
18. (a) $3, -2$ (b) $2.14, -2.81$
19. $\frac{116}{33}, -2$
20. $3, -2$
21. (a) $40° 5', 36° 8', 54° 31'$ (b) $x = 7, y = 3$
22. (a) 4 m, 7 m (b) $a = 2, b = 5, c = 4$
23. (a) $a = 4, b = 2, c = \frac{1}{2}$ (b) $x = 5, -1;\ y = 1, -1$
24. (a) $x = \frac{1}{2}, y = -1\frac{1}{2};\ x = 1\frac{2}{3}, y = 2$
    (b) $x = 11, y = 3;\ x = 2, y = -3$
    (c) $x = 3, y = \frac{1}{2};\ x = 6\frac{1}{2}, y = -\frac{17}{8}$
25. 7.5
26. $x = 1\frac{1}{2}$, 4 or -1
27. 2.78
28. 3.7
29. 87.0
30. $\frac{1}{3}, -1;\ 2, -2\frac{2}{3};\ 1\frac{1}{3}$
31. 2.68

Page 55 **EXERCISE 4(a)**

Self checking exercise. No answers provided.

Page 65 **EXERCISE 4(b)**

1. $x^4 - 2x^3 - 3x^2 - 4x$
2. 30.9936
3. 2.75

4. 1.32
5. 8.478; 0.9416; 0.2341; 0.002 646

(No answers provided for questions 6 to 10.)

Page 66 REVISION EXERCISE 4

1. $1 - x + x^2 - x^3 + x^4$
2. 0.768

(No answers are provided for questions 3 to 7.)

Page 74 EXERCISE 5(a)

1. 13 m; $A = 22° 37'$, $B = 67° 23'$
2. 18° 55½′; 37 m
6. 9 mm; 12° 41′

4. 12.5 m
5. 9.4 m
6. 520 mm
7. 110 m
8. 21° 48′
10. $\frac{2}{3}\sqrt{2}$; $\frac{1}{4}\sqrt{2}$
11. $\frac{80}{39}$, $\frac{80}{89}$, $\frac{89}{39}$
12. 1.94 m, 29°

Page 80 EXERCISE 5(b)

1. (a) 0.9604, 0.2787, 3.4456
   (b) 0.9843, − 0.1765, − 5.5753
   (c) − 0.3474, − 0.9377, 0.3706
   (d) − 0.4454, 0.8953, − 0.4975
2. (a) − 0.6428, − 0.7660, 0.8391
   (b) − 0.7580, 0.6157, − 1.2799
3. (a) 19° 28′ (b) 48° 11′ (c) 68° 12′
4. Maximum 1.414 when $x = 45°$; minimum − 1.414 when $x = 225°$
5. (a) 41° 49′, 138° 11′, 221° 49′, 318° 11′
   (b) 56° 47′, 123° 13′, 236° 47′, 303° 13′
   (c) 65° 54′, 114° 6′, 245° 54′, 294° 6′
6. (a) 0°, 66° 25′, 180°, 293° 35′, 360°
   (b) 48° 35′, 90°, 131° 25′, 270°
7. (a) 194° 29′, 210°, 330°, 345° 31′
   (b) 104° 29′, 180°, 255° 31′
   (c) 75° 58′, 135°, 255° 58′, 315°
8. (a) 45° (b) 105° (c) 216° (d) 900°

Page 81

REVISION EXERCISE 5

1. (b) (i) 45°, 153° 26′, 225°, 333° 26′
   (ii) 45°, 135°, 225°, 315°
2. (a) − 0.6534, − 1.3210 (b) 0.8090, − 1.7013
   (c) 0.9888, 6.687
3. (b) $-\tan x$
4. 2621 mm, 1836 mm, 1494 mm
5. Maximum 5 at $x = 36° 52'$
6. $a = 3.606$, $\theta = 56° 19'$
7. (a) (i) − 0.5736 (ii) 0.7660 (iii) − 0.3640
   (b) 14.5 m
8. 98° 25′, 148° 36′, 211° 24′, 261° 35′
9. 155.2 m; 38.6 m
10. 305 m
11. (a) 746 m (b) 1893 m
12. 41.7 m
13. (a) $\frac{5\pi}{4}$, 80° (b) 230°, 310° (c) $\frac{4}{5}$, $\frac{3}{4}$
14. (a) $\frac{12}{13}$, $\frac{5}{13}$ (b) 3
15. (a) 57° 18′ (b) 0.698 rad
16. Maxima: 4.343 at $x = 57° 2'$, 0.2667 at $x = 203° 12'$;
    Minima: − 0.2667 at $x = 156° 48'$, − 4.343 at $x = 302° 58'$
17. (b) − 2.824
18. (i) − 1 (ii) $-\frac{1}{2}$ (iii) $-\frac{1}{2}$
19. $\theta = 45°$
20. (a) 0 (b) $\frac{2xy}{x^2 + y^2}$; $\frac{2xy}{x^2 - y^2}$
21. (a) $-\frac{2}{13}$
22. (a) $2 + \frac{1}{2}\sqrt{3}$; $2\sqrt{2} - \frac{1}{2}$; $-2\sqrt{3} - \frac{1}{2}$; $-2\sqrt{3} + \frac{1}{2}$
24. (a) 60°, 120° (b) 45°
25. 36° 52′, 323° 8′; 66° 25′, 113° 35′

Page 86

EXERCISE 6(a)

1. 81p
2. 450 kg/m^3
3. 1680 kg
4. 39.24 kN/m^2
5. 6.12 kN

Page 90

EXERCISE 6(b)

1. 420 m^2; 37 m
2. 68 m^2
3. 8 rolls
4. 24350 mm^2
5. 14 100 mm^2
6. 69° or 111°
7. 2230 mm^2; 44.6 mm; 49.5 mm
8. 6612 m^2
9. 660 litres/sec
10. 13.2 m
11. 71° 37′
12. 103° 8′
13. 6.45 m^2
14. 63 m^2; 91 m^2
15. 6.411 m^2
16. (a) 17 m (b) 1.25 m
17. (a) 2868 mm^2 (b) 3054 m^2 (c) 31 230 mm^2

Page 91

REVISION EXERCISE 6

1. £9.60
2. 173 kg
4. 490 mm^2
5. 273 m^2
6. 7854 m^3
7. 167 minutes
8. (a) 812.5 mm (b) 0.54 m^2
9. 2630 litres
10. 4882 m^2
11. 180 litres
12. 4.24 mm

Page 96

EXERCISE 7

1. 4320 m^3
2. 543 m^3
3. 482 m^3
4. 75.6 m^3
5. 3.0 m^3
6. 355 m^3
7. 3070 m^3
8. 4744 m^3
9. 2175 m^3
10. 2140 m^3
11. 5.8 m^3

Page 98

REVISION EXERCISE 7

1. 95 m^3/s
2. (a) 77 m^3 (b) 79 m^3
3. Over 25 million tonnes
4. 5064 m^3
5. 47 m^3
7. 242 m^3
8. 885 m^3
9. 20 mm

Page 105 — EXERCISE 8(a)

1. 1.30 m; 1.54 m
2. 2.37 m
3. (a) 30° 49′ (b) 5 m
4. (a) 4.9 m (b) 93° 28′
5. 71°, 62°, 32.5
6. 726 m, 42° 8′, 29° 52′
7. 77 m

Page 107 — EXERCISE 8(b)

1. 115° 12′
2. 92° 8′; 27 m^2
3. 28° 4′, 53° 8′, 98° 48′; 84 m^2
4. 306 m^2; 83° 16′, 90°, 143° 8′, 43° 36′
5. 9010 m^2; 196 m
6. 126 m^2; 12 m

Page 107 — REVISION EXERCISE 8

1. $AB = 20$ m, $AC = 18$ m, $A = 12°\ 50'$, $C = 137°\ 10'$
2. 2.6 m; 52° 10′, 80° 50′
3. 31 178 m^2; 275 m
4. 2.4 km, 28° 22′
5. 98 m
6. 80° 40′; 71.6 m
7. 26 m; 73° 30′, 56° 27′, 50° 3′
8. 60 m
9. 126 m; 72° 45′, 78° 14′, 99° 1′
10. 83° 44′
11. (a) 6.46 m (b) 6.55 m
12. 129 m; 120 m
13. (a) 59 m; 97 m (b) 2790 m^2
14. 75.4 m; 82.2 m
15. 78° 28′, 44° 25′, 57° 7′; 1.4 m; 2.22 m^2
16. (a) 1.04 m
    (b) 85° 47′, 62° 22′
    (*Note*: Values obtained by the sine rule are likely to be inaccurate in this question as angle DBA is close to 90°.)
17. 485 m
18. 75 m
19. 2213 m; 798.5 m; 685.4 m
20. 669 mm; 553 mm

Page 114 — EXERCISE 9(a)

1. $3x^2$
2. 3; 12; 27; 13
3. (a) $5x^4$ (b) $3.5x^{2.5}$ (c) $\frac{3}{4}x^{-1/4}$
   (d) $-28x^{-5}$ (e) $\frac{5}{2}x^{-3/8}$ (f) $-\frac{3}{4}x^{-1/8}$
   (g) $-2x^{-7}$ (h) $\frac{7}{16}x^{-1/2}$ (i) $-0.18x^{-1.02}$
   (j) $\frac{1}{n}x^{\frac{1-n}{n}}$

4. (a) $-\frac{9}{x^{10}}$ (b) $-\frac{1}{2\sqrt{x^3}}$ (c) $-\frac{1}{3x^2}$

(d) $-\frac{4}{5x^5}$ (e) $\frac{1}{\sqrt{x}}$ (f) $\frac{1}{\sqrt{2x}}$

5. (a) $6x^2 + 6x - 4$ (b) $3\sqrt{x} - \frac{2}{\sqrt{x}}$

6. (a) $10(5x - 3)$ (b) $-3(2 - x)^2$ (c) $\frac{-3}{2\sqrt{(4 - 3x)}}$

(d) $\frac{1}{(5 - x)^2}$ (e) $\frac{-56}{(4x - 3)^3}$ (f) $\frac{6}{\sqrt{(1 - 3x)^3}}$

7. $\sqrt{\frac{a}{x}}$; $\sqrt{a}$, 1, $\frac{1}{2}$; $x = \frac{1}{2\sqrt{a}}$

8. (a) $6x^2(x^3 - 4)$ (b) $3(1 + 2x)(1 + x + x^2)^2$

(c) $\frac{-48x}{(6x^2 + 5)^5}$ (d) $\frac{6}{\sqrt{x}} + 6 + \frac{3\sqrt{x}}{2}$ (e) $\frac{x - 1}{\sqrt{(x^2 - 2x)}}$

Page 118 EXERCISE 9(b)

1. $6x$

2. $-\frac{l}{\theta^2}$

4. 25 s

5. $6x - \frac{2}{(1 - x)^2}$

6. (a) $9(2 - 5x)(2 - x)^3$ (b) $\frac{x(2 + 5x)}{\sqrt{(1 + 2x)}}$

(c) $\frac{3x^2(8 - 7x)}{2\sqrt{(4 - 3x)}}$ (d) $\frac{12x - 1}{(5 - 4x)^2}$

(e) $\frac{x(16 - 3x)}{2(4 - x)^{3/2}}$

Page 121 EXERCISE 9(c)

(*Note*: Arbitrary constants have been omitted)

1. (a) $2x^3$ (b) $2x^4$

(c) $\frac{2}{3}x^{3/2}$ (d) $\frac{x^{1.15}}{1.15}$

2. $y = 3x^3 + 3x^2 - 10x + 4$

3. $s = 112.5$

4. (a) $-\frac{1}{2x}$ (b) $\frac{-2}{3x^{1/5}}$

(c) $8\sqrt{x}$ (d) $-\frac{(x + 2)}{2x^3}$

5. (a) $\frac{1}{8}(3 - 2x)^4$ (b) $\frac{1}{6}(1 + 4x)^{3/2}$

6. (a) $\frac{4}{3}\pi r^3$ (b) $300V^{2.4}$ (c) $4t - \frac{4}{15}t^3$

7. (a) $833\frac{1}{3}$ (b) $\frac{18}{81}$

Page 122

REVISION EXERCISE 9

1. (a) $\frac{ds}{dt} = 64$ (b) (i) $\frac{9}{2}\sqrt{t}$ (ii) $9(3x + 2)$

   (c) $-3;\ 2;\ -6$

2. (a) $24x$

   (b) (i) $10x^4 - \frac{1}{4\sqrt{x}} - \frac{1}{2x^{3/2}}$

   (ii) $\frac{x^2 - 8x - 3}{(x - 4)^2}$

   (iii) $\frac{5x^2 + 4x + 3}{2\sqrt{(x + 1)}}$

3. (a) (i) $-4/x^2$ (ii) $18x^5$ (iii) $80x^9$

   (iv) $3x^2$ (v) $7 - 16x$

   (b) (i) $3x^2 + 2x + C$ (ii) $3x^3 + \frac{4}{3}x^6 + C$ (iii) $\frac{7}{2}x^2 + C$

   (iv) $\frac{9}{2}x^2 + 18x + C$ (v) $\frac{x^{11}}{11}$

4. (i) $-\frac{1}{2\sqrt{(1 - x)}}$ (ii) $-\frac{3}{2x^4}$ (iii) $\frac{4}{(1 - 3x)^2}$

5. (a) $-\frac{8.2}{3}$ (b) $-\frac{8.02}{3};\ -\frac{8}{3}$

6. (a) 3.05 (b) 3.005; 3.000

7. (b) (i) $12x^3 + \frac{5}{x^6}$ (ii) $(2x - 1)(1 + x)^2x^{-2}$

   (iii) $-\sqrt{1.92x^{-2.2}}$

8. (a) (i) $-\frac{1}{6}x^{-6} - 20\sqrt{x} + 5x$ (ii) $10\sqrt{3}x^{0.1}$

   (iii) $x^3 + \frac{1}{5}x^{5/4}$

   (b) $y = \frac{1}{6}(x^6 - 2x^3 + 1);\ 0, 2^{2/3}$

Page 131

EXERCISE 10

1. 90 m
2. 14 m/s
3. 62.5 m
4. 86.4 m
5. 7 s, 2 rad
6. 9
7. Max $(-2, 24)$, Min $(4, -60)$
8. (a) Max $(-2, -4)$, Min $(2, 4)$
   (b) Min $(\frac{1}{3}, 12)$
   (c) Max $(2, \frac{5}{4})$

9. 5, 2, -3; Max $(\frac{2}{3}, 21\frac{5}{7})$, Min $(-4, -54)$

10. (a) Max (2, 13) (b) Min $(-1, -34)$, Max $(\frac{1}{2}, 4\frac{1}{8})$, Min $(2\frac{1}{2}, -76\frac{7}{8})$

11. 50 mm

12. 2:1

13. $\pi/6$

14. 3 m

Page 131 REVISION EXERCISE 10

1. (a) 250 units^3
   (b) $\frac{1}{2}\pi$ s, π s; $\frac{1}{4}\pi$ s, $\frac{3}{4}\pi$ s; 20 m/s, $-$ 20 m/s

2. (a) $4x^3 - 3x^2 - 8x + 19$; $\dfrac{17}{(2x+3)^2}$; $36x^2(2x^3-3)^5$
   (b) Max (1, 11), Min (2, 10); 12

3. (a) $-3/6\pi^2x^2$; $1.2x^{-3/2} - 0.16x^{-1/2}$; $18x + \dfrac{2}{9x^3}$
   (b) 19.2 m/s; 14.4 m

4. $\dfrac{x}{\sqrt{(1+x^2)}}$; $-\dfrac{12}{x^5}$; $\dfrac{2x-x^2}{(1-x)^2}$
   Max (0, 1), Min (2, -7)

5. (a) -3, 2
   (b) $22\frac{1}{2}$, $1\frac{3}{8}$

6. 4 m from A; 96 mm

7. (a) + 1 or $-$ 1
   (b) $10(2x+3)^4$; $\dfrac{-2}{\sqrt[3]{x^5}}$
   (c) 20 mm squares

8. (a) $-\dfrac{2}{3x^3}$; $\dfrac{1}{\sqrt{1+x^2}}$; $\dfrac{1}{(1+2x)^2}$
   (b) 20 mm by 10 mm

9. (a) $\dfrac{1+x}{1-x}$; $-\dfrac{5}{2x^6}$; $-\dfrac{1}{(1-3x)^{2/3}}$
   (b) 30, $-11\frac{2}{3}$

10. (a) $4(2-3x-6x^2)$; $6x(x-5)/(2x-5)^2$
    (b) 2 m; 12 m^2

11. (a) $3(7x+1)(x+2)^6$; $3/(x+1)^2$
    (b) height : diameter = 1:1

12. (a) 14.16 (b) 9.45

13. $2lw/3EI$; $2lw/3$

14. (a) $\frac{1}{2}$ (Minimum), -2 (Maximum) (b) 26 mm by 39 mm

---

Page 138

EXERCISE 11

1. 625; 18; 3
2. (a) 1/6 (b) 8/3
3. (a) $3\frac{1}{3}$; $\frac{1}{6}$ (b) $x - \frac{2}{3}x^3 + \frac{1}{5}x^5 + C$; $\frac{1}{4}(x^2 + 1)^2 + C$
4. 67.2 units^2
5. 36 unit^2
6. 6; 8π
7. $y = \frac{I}{EI}\left[2x^3 - \frac{x^4}{12} - 144x\right]$
8. $y = \frac{w}{24EI}[6L^2x^2 - 4Lx^3 + x^4]$; $\frac{wL^4}{8EI}$; $\frac{wL^3}{6EI}$

Page 138

REVISION EXERCISE 11

(Note that arbitrary constants have been omitted)

1. (a) $\frac{2}{5}\sqrt{x^5} + 4\sqrt{x} + \frac{2}{3}\sqrt{x^3}$
   (b) $17a^2/6$
2. 0.605; 0.758; 22.8
3. $10\frac{2}{3}$ units^2
4. 24 units^2
5. 36 units^2
6. (a) $-\frac{1}{4(4x + 3)}$ (b) 1.4
7. (a) $\frac{1}{5}x^5 - x^3 - 5x$; $(3t^2 - 2)^9$; $-\frac{3}{2x^2}$ (b) 18π units^3
8. $\frac{1}{2}x^2 + \frac{1}{x}$; $\frac{8}{3}\sqrt{x^3}$; $x^3 - 2x^2 + 3x - 1$; $x = ut + \frac{1}{2}at^2$

10. $y = x^3 - 6x^2 + 24x + 12$

Page 144

EXERCISE 12(a)

2. $\frac{1}{2}$; $\frac{3}{32}$; $\frac{5}{16}$
3. 1/270 725; 11/4165; 99/54 145
4. $\frac{1}{25}$; $\frac{1}{5}$; $\frac{1}{5}$
5. $\frac{1}{2}$; $\frac{1}{2}$
6. 1/506
8. 30
9. 15 600; 17 576
10. 1/625

Page 149

EXERCISE 12(b)

1. 1/12; 5/12
2. 1/80
3. 253/300; 1/300; 46/300
4. 0.06; 0.56; 0.24
5. 0.145
6. 1/192; 35/64

Page 149

REVISION EXERCISE 12

1. 105/703
2. 1/12; 1/365
3. 48
4. 0.70; 0.0096; 0.088
5. 0.038; 0.048; 0.02; 0.088; 0.912
6. 1/36
7. 1/969; 32/323; 364/969; 33/323
8. (a) 3; 12; 8 (b) $\frac{1}{2}$; $\frac{1}{8}$; $\frac{1}{8}$
9. 11/36
10. 8/150
11. 0.45; 0.10; 0.90
12. 1/560
13. 2/15
14. 34/1265

Page 162

EXERCISE 13(a)

2. Mode is 2 children, mean is nearer to 3 children.
3. 48.8 marks.

Page 173

EXERCISE 13(b)

1. 690 mm; 20.4 mm
2. 10.5%; 1.56%
3. 44.6 hrs; 45 hrs; 1.9 hrs
4. £30.4; £3.76
5. 22.9 N/mm^2; 2.37 N/mm^2
6. 1170 mm; 24.6 mm
7. 51.2 marks; 20 marks
8. 54.5; 10
9. 0.092
10. 24 MPa; 25.3 MPa; 2.9 MPa

Page 175

REVISION EXERCISE 13

2. Mean is 8.39 min; median is in the 8–8½ group at 8.4 approx.; modal group is the 8–8½ group.
3. 6.05 mm; 6.04 mm; 6.00 mm; 0.305 mm.
4. (a) 67° 30′ 14.8″
   (b) 4, 10, 2
5. 29%; 2¼%
6. 3.5 men; 3 men; 4.1 men; 1.7 men
7. 14.4 N/mm^2
8. £24; £23; £22; £3.09
9. 42.6 hours; 43 hours; 1.9 hours
10. 13.5 min; 1.965 min; 2.65 min
11. (a) 17° 5′, 21° 32′, 141° 23′ (b) 11.36 mm
12. 0.8667; 6.7%
13. (a) 10.0, 2.0 (b) 1900 (95.4%)